Student's Solutions Manual

Math Reasoning for Elementary Teachers

Calvin Long
Duane W. DeTemple

Washington State University

An imprint of Addison Wesley Longman, Inc.

Reading, Massachusetts • Menlo Park, California • New York • Harlow, England
Don Mills, Ontario • Sydney • Mexico City • Madrid • Amsterdam

Reproduced by Addison-Wesley from camera-ready copy supplied by the authors.

Copyright © 1996 Addison-Wesley Educational Publishers, Inc.

All rights reserved. No part of this publication may be reproduced, stored in a retrieval system, or transmitted, in any form or by any means, electronic, mechanical, photocopying, recording, or otherwise, without the prior written permission of the publishers. Printed in the United States of America.

ISBN   0-673-99391-4

2 3 4 5 6 7 8 9 10 CRS 99

Table of Contents

Chapter 1	Thinking Critically	1
Chapter 2	Sets, Whole Numbers, and Functions	17
Chapter 3	Numeration and Computation	37
Chapter 4	Number Theory	54
Chapter 5	Integers	71
Chapter 6	Fractions and Rational Numbers	81
Chapter 7	Decimals and Real Numbers	95
Chapter 8	Statistics: The Interpretation of Data	105
Chapter 9	Probability	116
Chapter 10	Geometric Figures	132
Chapter 11	Congruence, Constructions, and Similarity	151
Chapter 12	Measurement	161
Chapter 13	Geometric Transformations	172
Chapter 14	Coordinate Geometry	184
Appendix B	Logic and Mathematical Reasoning	200
Appendix C	Logo: A Programming Language for Learning Mathematics	202

Chapter 1

Problem Set 1.1 (page 7)

1. (a)
$$9 \times 9 = 81$$
$$79 \times 9 = 711$$
$$679 \times 9 = 6111$$
$$5679 \times 9 = 51,111$$
$$45,679 \times 9 = 411,111$$
$$345,679 \times 9 = 3,111,111$$
$$2,345,679 \times 9 = 21,111,111$$
$$12,345,679 \times 9 = 111,111,111$$

3. (a)
$$1 \times 8 + 1 = 8 + 1 = 9$$
$$12 \times 8 + 2 = 96 + 2 = 98$$
$$123 \times 8 + 3 = 984 + 3 = 987$$

5. (a)
$$67 \times 67 = 4489$$
$$667 \times 667 = 444,889$$
$$6667 \times 6667 = 44,448,889$$

6. (b) Each result is 5. For example, in the top row, $6 + 8 - 9 = 5$

6	9	8
3	5	7
2	1	4

9. (a)
$$1 \times 142,857 = 142,857$$
$$2 \times 142,857 = 285,714$$
$$3 \times 142,857 = 428,571$$
$$4 \times 142,857 = 571,428$$
$$5 \times 142,857 = 714,285$$

11. (a)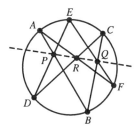

The points P, Q, and R appear to be collinear, that is, they lie on the same line.

13. (a)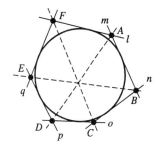

$\overline{AD}$, $\overline{BE}$, and $\overline{CF}$ appear to intersect at a point, that is, they are concurrent.

18. (a) Yes, it appears very likely. It has been checked by computer for starting values from 1 to well up into the millions. However, no one has been able to prove that the process will always terminate. Starting with 7 we generate the following:
$$3 \times 7 + 1 = 22$$
$$22 \div 2 = 11$$
$$3 \times 11 + 1 = 34$$
$$34 \div 2 = 17$$
$$3 \times 17 + 1 = 52$$
$$52 \div 2 = 26$$
$$26 \div 2 = 13$$
$$3 \times 13 + 1 = 40$$
$$40 \div 2 = 20$$
$$20 \div 2 = 10$$
$$10 \div 2 = 5$$
$$3 \times 5 + 1 = 16$$
$$16 \div 2 = 8$$
$$8 \div 2 = 4$$
$$4 \div 2 = 2$$
$$2 \div 2 = 1$$

JUST FOR FUN   **For Careful Readers (page 13)**

1. The second engineer was the first engineer's daughter.

2. There was no smoke since the train was electric.

3. The 12th rung down since the ladder and the ship both rise with the water level.

Problem Set 1.2 (page 15)

1. (a) Using guess and check:
   Guess 14 bikes, 13 trikes. The number of wheels is $14 \times 2 + 13 \times 3 = 67$.
   Too many wheels, too many trikes.
   Guess again: 17 bikes, 10 trikes. The number of wheels is $17 \times 2 + 10 \times 3 = 64$.
   Still too many wheels.
   Guess again: 21 bikes, 6 trikes. The number of wheels is: $21 \times 2 + 6 \times 3 = 60$.
   O.K.

4. Work backwards from 52.
   Add 8: $8 + 52 = 60$.
   Divide by 5: $60 \div 5 = 12$.
   Guess and check or a carefully structured guess and check strategy, as in 1(b) above, also work well.

7. (a) The only possible sums of 3 digits totaling 19 are:
   $4 + 7 + 8 = 19$
   $4 + 6 + 9 = 19$.
   Since 4 is used twice, it must be in the middle.

9. (a) Answers will vary. Two possibilities are

   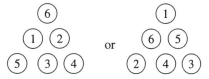

   Another possibility is given in part (c).

11. (d) Note that in each of (a), (b), and (c), the sum of the top and bottom numbers given is the sum of the left and right numbers; i.e., $11 + 20 = 15 + 16$, $10 + 21 = 12 + 19$, and $9 + 16 = 7 + 18$. For such a problem to have a solution, this must always be the case. Thus, (d) has no solution.

12. (a) Since $2 + 3 = 5$, $3 + 5 = 8$, $5 + 8 = 13$, and $8 + 13 = 21$, the sequence is 1, 2, 3, 5, 8, 13, 21.

(c) This can be solved using the guess and check strategy. A more formal solution is as follows:
Suppose the second number is n. Then the third number is $3 + n$, and the fourth number is 13. Therefore, $n + (3 + n) = 13$, so $n = 5$. (Note that the guess and check strategy can also be used). The sequence is 3, 5, 8, 13, 21, 34, 55.

(e) Suppose the second number is n. Then the sequence is 2, n, $2 + n$, $2 + 2n$, $4 + 3n$, 11. Therefore $(2 + 2n) + (4 + 3n) = 11$, so $6 + 5n = 11$ and $n = 1$. The sequence is 2, 1, 3, 4, 7, 11.

Answer: 8 nickels and 7 dimes.

JUST FOR FUN   How Many Heaps? (page 24)

Make an orderly list:

Number of fruits in each heap	Total number of fruits
1	$(1 \times 63) + 7 = 70$
2	$(2 \times 63) + 7 = 133$
3	$(3 \times 63) + 7 = 196$
4	$(4 \times 63) + 7 = 259$
5	$(5 \times 63) + 7 = 322$

Since 322 is the first number on the right that is divisible by 23, and $322 \div 23 = 14$, the least number of fruits each traveler could have received is 14.

Problem Set 1.3 (page 25)

1. No, because $5 \times 10 + 13 = 63$.

3. Using the guess and check and making an orderly list of strategies, we construct the following table.

Guess for Lisa's number	Number that results
1	$7 \cdot 1 - 4 = 3$
2	$7 \cdot 2 - 4 = 10$
3	$7 \cdot 3 - 4 = 17$

Lisa's number is 3. Suppose, on the other hand, that we wanted to obtain 66 as the result in this problem. Since $66 - 17 = 49$ and $49 \div 7 = 7$, we would have to proceed seven more steps. Jumping seven steps from 3 to 10 we obtain $7 \cdot 10 - 4 = 66$ as desired.

5. (a) Yes. Yes. Juan's rule could have been either one of these. The result obtained by the first rule is $5n - 3$. The result obtained by the second rule is
$5(n - 1) + 2 = 5n - 5 + 2 = 5n - 3$,
which is the same as the first rule. Without algebra, you can reason that multiplying one less than a number by five gives a result that is 5 less than the original number multiplied by 5. Then adding 2 gives a number which is 3 less than the original number multiplied by 5.

7. (a) Add: $7 + 1 = 8, 1 + 4 = 5, 4 + 7 = 11$.

(c) Use the method shown on page 21. The sum of the three new numbers must be $\frac{22+19+17}{2} = 29$. Note that
$29 - 22 = 7, 29 - 19 = 10$, and
$29 - 17 = 12$.

(e) The sum of the three new numbers must be $\frac{7+11+13}{2} = 15.5$. Note that
$15.5 - 7 = 8.5, 15.5 - 11 = 4.5$, and
$15.5 - 13 = 2.5$.

9. There are 49 ways, listed below in order of number of pennies used. For example, Q + D + 15P means a quarter, a dime, and 15 pennies.

2Q	Q + D + 15P
Q + 2D + N	Q + 2N + 15P
Q + D + 3N	3D + N + 15P
Q + 5N	2D + 3N + 15P
5D	D + 5N + 15P
4D + 2N	7N + 15P
3D + 4N	Q + N + 20P
2D + 6N	3D + 20P
D + 8N	2D + 2N + 20P
10N	D + 4N + 20P
Q + 2D + 5P	6N + 20P
Q + D + 2N + 5P	Q + 25P
Q + 4N + 5P	2D + N + 25P
4D + N + 5P	D + 3N + 25P
3D + 3N + 5P	5N + 25P
2D + 5N + 5P	2D + 30P
D + 7N + 5P	D + 2N + 30P
9N + 5P	4N + 30P
Q + D + N + 10P	D + N + 35P
Q + 3N + 10P	3N + 35P
4D + 10P	D + 40P
3D + 2N + 10P	2N + 40P
2D + 4N + 10P	N + 45P
D + 6N + 10P	50P
8N + 10P	

11.

1357	3157	5137	7135
1375	3175	5173	7153
1537	3517	5317	7351
1573	3571	5371	7315
1735	3715	5713	7513
1753	3751	5731	7531

13. Use the strategy of making an orderly list. Assume that the pearls and the bags are identical, so that, for example, 21, 1, 3 is considered the same as 21, 3, 1. Then we need list only the possibilities in which the number of pearls in bag 1 is at least as great as the number of pearls in bag 2, and the number of pearls in bag 2 is at least as great as the number of pearls in bag 3. We have the following:

Bag 1	Bag 2	Bag 3
23	1	1
21	3	1
19	3	3
19	5	1
17	5	3
17	7	1
15	9	1
15	7	3
15	5	5
13	9	3
13	7	5
13	11	1
11	7	7
11	9	5
11	11	3
9	9	7

15. Make an orderly list.

Number pair	Sum
4, 24	28
6, 16	22
8, 12	20

17. A diagram of the situation will show that Bob has to make 9 cuts to get 10 2-foot sections. Since each cut takes one minute, it will take Bob 9 minutes to do this.

20. Though it may seem unusual, the simplest approach to this problem is that of drawing a diagram as in Example 1.5. Make the analogy between the race in Example 1.5 and the political race in the present problem. Draw a line with equally spaced points with the distance between consecutive points representing 1000 votes. Then place A (for Albright), B, C, D, and E on the line according to the statement of the problem.

 B   A   E C D

Place A at an arbitrary point on the line and B two units to A's left since Albright finished 2000 votes ahead of Badgett, and so on. The completed diagram is as shown and the order of finishing from first to last is Dawkins, Chalmers, Ertl, Albright, and Badgett.

22. The total area of the lawn plus walkway is $11 \times 14 = 154$ square meters. The area of the lawn alone is $9 \times 12 = 108$ square meters. Therefore, the area of the walkway is $154 - 108 = 46$ square meters.

Problem Set 1.4 (page 41)

1. (a) 2, 5, 8, 11, 14, 17, 20. Each succeeding term is 3 more than the preceding term.

   (c) 1, 1, 3, 3, 6, 6, 10, 10, 15, 15. The sequence consists of the triangular numbers repeated twice.

   (e) 2, 6, 18, 54, 162, 486, 1458. Each term is three times the preceding term.

2. (a) Each pattern has one more column of dots than the previous one. The next three patterns are shown.

   (d) The nth even number

   (f) Using Gauss' method:
   $$s = 2 + 4 + 6 + \cdots + 2402$$
   $$s = 2402 + 2400 + 2398 + \cdots + 2$$
   $$2s = 2404 + 2404 + 2404 + \cdots + 2402$$
   There are 1201 terms of 2404. Thus, the sum of the series is
   $2,887,204 \div 2 = 1,443,602$.

4. (a) Each term is 2 more than its predecessor. Since $35 = 5 + 30$ and $30 \div 2 = 15$, we conclude that 15 2s must be added to 5 to get 35. Thus, there are 15 terms after the first one, or 16 terms in all.

5. (a) The terms form the sequence in Problem 4(a), so there are 16 terms. Use Gauss's method.
   $$s = 5 + 7 + 9 + \cdots + 35$$
   $$s = 35 + 33 + 31 + \cdots + 5$$
   $$2s = 40 + 40 + 40 + \cdots + 40$$
   $2s = 16 \times 40 = 640$, so the sum is 320.

   (c) The terms form the sequence in Problem 4(c), so there are 17 terms. Use Gauss's method.
   $$s = 3 + 7 + 11 + \cdots + 67$$
   $$s = 67 + 63 + 59 + \cdots + 3$$
   $$2s = 70 + 70 + 70 + \cdots + 70$$
   $2s = 17 \times 70 = 1190$, so the sum is 595.

7. (a) The middle term on the left side of the nth equation is n. The next two lines are:
$1 + 2 + 3 + 4 + 5 + 4 + 3 + 2 + 1 = 25$
$1 + 2 + 3 + 4 + 5 + 6 + 5 + 4 + 3 + 2 + 1 = 36$.

9. Notice the pattern: Except for the last column of the table, when you move right one square, the units digit increases by 1, and when you move down one square, the tens digit increases by 1. Hence, we fill in the arrays and determine the desired last number as shown.

(a)

53	54	55	56
			66
			76
			86

The desired number is 86.

(d)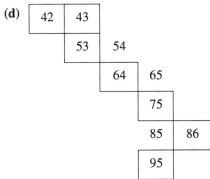

The desired number is 42.

11. (a) The left side of the equation has an additional square number added or subtracted in each step. Additions and subtractions alternate.
$1 - 4 + 9 - 16 + 25 = 15$
$1 - 4 + 9 - 16 + 25 - 36 = -21$

(b) The sequence 1, 3, 6, 10, ..., is the sequence of triangular numbers, $t_n = \frac{n(n+1)}{2}$ (see page 34). Since the right sides of the equation alternate between positive and negative, we expect the seventh equation to have $t_7 = \frac{7 \cdot 8}{2} = 28$ on the right, and the eighth equation should have $-t_8 = -\frac{8 \cdot 9}{2} = -36$. The equations are:

$1 - 4 + 9 - 16 + 25 - 36 + 49 = 28$
$1 - 4 + 9 - 16 + 25 - 36 + 49 - 64 = -36$

(c) n even: $1 - 4 + 9 - \cdots - n^2 = -\frac{n(n+1)}{2}$

n odd: $1 - 4 + 9 - \cdots + n^2 = \frac{n(n+1)}{2}$

14. (a) As shown in the diagram in the text, there are 6 segments.

(c) Each dot is connected to 99 other dots, so 99 segments end at each dot. By multiplying 100×99, we count each segment twice (since a segment has 2 endpoints), so the number of segments is $\frac{100 \times 99}{2} = 4950$.

17. (a) In each figure, dots are added to the upper left, upper right, and lower right sides to complete the next larger pentagon.

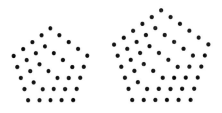

(b) 1, 5, 12, 35, 51, ...

(c) $1 + 4 + 7 + 10 + 13 = 35$
$1 + 4 + 7 + 10 + 13 + 16 = 51$

(d) 10th term = $1 + 3(9) = 28$

(e) Use Gauss's method:
$s = 1 + 4 + 7 + \cdots + 28$
$s = 28 + 25 + 22 + \cdots + 1$
$2s = 29 + 29 + 29 + \cdots + 29$
Sum $= \frac{(10)(29)}{2} = 145$

(f) nth term $= 1 + 3(n - 1) = 3n - 2$

(g) Using Gauss's method, there are n terms of $(3n - 1)$. The sum is $\frac{n(3n-1)}{2}$.
Therefore, $p_n = \frac{n(3n-1)}{2}$.

21. (a) Row 1: 1
Row 2: 2
Row 3: 3 + 1 = 4
Row 4: 4 + 4 = 8
Row 5: 5 + 10 + 1 = 16
Row 6: 6 + 20 + 6 = 32
Row 7: 7 + 35 + 21 + 1 = 64
Row 8: 8 + 56 + 56 + 8 = 128

22. (a) Row 0: $1^2 = 1$
Row 1: $1^2 + 1^2 = 2$
Row 2: $1^2 + 2^2 + 1^2 = 6$
Row 3: $1^2 + 3^2 + 3^2 + 1^2 = 20$
Row 4: $1^2 + 4^2 + 6^2 + 4^2 + 1^2 = 70$

(b) The sums are the entries of the vertical column in the middle of the Pascal triangle.

23. (a) For 4: $\sqrt{1 \cdot 1 \cdot 3 \cdot 6 \cdot 10 \cdot 5} = \sqrt{900} = 30$.
For 15: $\sqrt{6 \cdot 5 \cdot 10 \cdot 20 \cdot 35 \cdot 21}$
$= \sqrt{4,410,000} = 2100$.
For 35: $\sqrt{21 \cdot 15 \cdot 20 \cdot 35 \cdot 70 \cdot 56}$
$= \sqrt{864,360,000} = 29,400$.

JUST FOR FUN   **How Many Pages in the Book? (page 49)**

To number pages 1 through 9 takes 9 · 1 = 9 digits. To number pages 10 through 99 takes 90 · 2 = 180 digits. This leaves 867 − 180 − 9 = 678 to number 3-digit pages. Thus, there are 678 ÷ 3 = 226 3-digit pages and 226 + 90 + 9 = 325 pages in the book.

Problem Set 1.5 (page 54)

1. The second player can always add a sufficient number of tallies to reach a multiple of 5 at each step. This will force the first player to go over 30.

3. (a) Work backwards. Before the last jump, Josh had $16, since $16 \times 2 = 32$. Before the second jump, Josh had $\frac{1}{2}(16+32) = \$24$. Before the first jump, Josh had $\frac{1}{2}(24+32) = \$28$. He started with $28.

6. Coats: Since Joe was wearing Moe's coat, Hiram must have been wearing Joe's coat. Therefore, Moe was wearing Hiram's coat.
Hats: Since Joe was wearing Hiram's hat, Moe must have been wearing Joe's hat. Therefore, Hiram was wearing Moe's hat.

7. (a) Since the number is greater than 20, less than 35, and divisible by 5, it must be either 25 or 30. Since the sum of the digits is 7, it must be 25. Not all information was needed—for example, we did not use the first and second clues.

8. Since Beth and the guard bought something for Mitzi, neither Beth nor Mitzi is the guard—so Jane is the guard. Since Beth is neither the guard nor the forward, she is the center. This leaves Mitzi to be the forward.
Conclusion: Beth is the center, Jane is the guard, Mitzi is the forward.

11. Yes. The problem can be solved with 5 or 7 in the center. On the other hand, it cannot be solved with 6 or 8 in the center because in both cases one of the two remaining sums must be odd and the other even.

12. (a) 3. Use the pigeonhole principle—in this case, the children are the "pigeons" and genders are the "holes."

14. If the difference $a - b$ is divisible by 10, that means that $a - b$ is a multiple of 10. In any set of 11 natural numbers, at least two of the numbers must have the same units digit which implies that their difference is a multiple of 10.

16.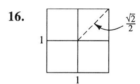

Using the pigeonhole principle where the "pigeons" are the five points and the "pigeonholes" are the small squares with their boundaries (see note below), there must be two points that are in the same pigeonhole or small square. The farthest apart that these two points can be is the length of the diagonal of the small square, or $\frac{\sqrt{2}}{2}$. Therefore, there are at least two points that are at most $\frac{\sqrt{2}}{2}$ units apart.

Note: The pigeonholes as described above overlap at some of the boundaries. To be precise, the pigeonholes should not overlap. This can be accomplished by excluding the left boundaries from the pigeonholes on the right, and excluding the top boundaries from the bottom pigeonholes.

18. The total number of marbles is
$1+2+3+\cdots+10 = \frac{10 \cdot 11}{2} = 55$. There are 10 groups of 3 adjacent cups of marbles. If we add the number of marbles in each group, then each marble is counted 3 times, so the new total is $3 \cdot 55 = 165$. If each group had 16 or fewer marbles, the total would be at most $10 \cdot 16 = 160$ (too small). Therefore, at least one group of three adjacent cups must contain at least 17 marbles.

20. The number of people at the party with no friends is none, exactly one, or 2 or more.

    Case (i): If everyone has at least one friend, then each of the 20 people at the party has between 1 and 19 friends, inclusive. By the pigeonhole principle, at least two of them have the same number of friends.

    Case (ii): If exactly one person has no friends, then each of the other 19 people has 1 to 18 friends at the party. By the pigeonhole principle, at least two of them have the same number of friends.

    Case (iii): If 2 or more people have no friends, then they have the same number of friends at the party.

Problem Set 1.6 (page 65)

1. Work backwards as shown.

    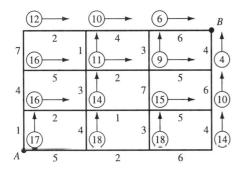

    The least time is 17 minutes. Each of the following routes produces the optimal time: NENENE; NENNEE; NENEEN.

5. Sarah gets the racket. She pays Ken $11.25 because the average value is $22.50 and half of this is $11.25.

9. (a) The upper right circle contains an odd number of 1s, so we correct it as shown.

    1 1 1 0 1 0 0

   (c) Correct. Each circle has an even number of 1s.

10. (a) The upper left circle has an odd number of 1s, so we correct it as shown.

    1 1 0 1 0 0 0

    (b) The single error must be in the intersection of all three circles. Changing the bit in this intersection transforms all "bad" circles into "good" circles.

13. (a) Since the stores do the same amount of business, the cost of supplying them depends on their total distance from the warehouse. We use a variable and try the various cases.

    (i) $\underset{\underset{W}{d}}{\underline{\overset{A \quad 4 \quad B \; 2 \; C}{\bullet \qquad \bullet \; \bullet}}}$

    Suppose the warehouse is d miles from A toward B. Then the total distance is $d + (4 - d) + (6 - d) = 10 - d$. This is smallest when d is largest; i.e. when $d = 4$, $10 - d = 6$, and the warehouse is at B.

    (ii) $\underset{\underset{d \; W}{}}{\underline{\overset{A \quad 4 \quad B \; 2 \; C}{\bullet \qquad \bullet \; \bullet}}}$

    Suppose the warehouse is d miles from B toward C. Then the total distance is $(4 + d) + d + (2 - d) = 6 + d$ and this is least when $d = 0$, $6 + d = 0$, and warehouse is at B.
    Thus, the warehouse should be located at B.

CLASSIC CONUNDRUM   10 = 9? (page 69)

No, no. The two men in the first room were number 1 and number 2. The proprietor's scheme actually ignores the tenth man entirely!

Chapter Review Exercises (page 72)

1. (a) $1 \cdot 8 = 8$
$21 \cdot 8 = 168$
$321 \cdot 8 = 2568$
$4321 \cdot 8 = 34,568$
$54,321 \cdot 8 = 434,568$
$654,321 \cdot 8 = 5,234,568$
$7,654,321 \cdot 8 = 61,234,568$
$87,654,321 \cdot 8 = 701,234,568$
$987,654,321 \cdot 8 = 7,901,234,568$

   (b) No. There is a pattern. The first digit increases sequentially from 0 to 7, except for the last product (also, think of 8 as 08). The remaining string of digits has the pattern 8, 68, 568, 4568, 34,568, ...

   (c) Yes.

   (d) No. Patterns may break down at a certain point. To make a generalization requires mathematical proof.

2. (a) Answers will vary. For example, suppose you choose "5" for a favorite digit. Performing the indicated multiplications produces a string of 7 fives. This string of seven of your favorite digit always occurs irrespective of the choice of digit.

   (b) 1,111,111

3. (a) The sum in each row, diagonal, and column must be
$$\frac{1+7+13+31+37+43+61+67+73}{3}$$
= 111. One possibility is shown.

67	1	43
13	37	61
31	73	7

   (b) Answers will vary. One way to get a magic subtraction square is to interchange the "end" numbers of each diagonal—in this case, 67 and 7, and 31 and 43.

7	1	31
13	37	61
43	73	67

4. *P*, *Q*, and *R* appear to be collinear, that is, they lie on the same line. See the illustration in the text.

5. 87 + 78 = 165, 165 + 561 = 726, 726 + 627 = 1353, 1353 + 3531 = 4884 which is a palindrome.

6. (a) Methods will vary. Use trial and error and make a chart similar to the one below:

Number of 8' boards	Number of 10' boards	Total number of feet
45	45	810 (too small)
40	50	820 (too small)
35	55	830 (too small)
30	60	840 (too small)
28	62	844 (o.k.)

   Therefore, there were 28 eight-foot boards.

   (b) Methods will vary. Use Jennifer's method. If all boards were 8 feet long, the total length would be $90 \cdot 8 = 720$. He has $844 - 720 = 124$ "extra" feet. Since each 10 foot board has 2 "extra" feet, the number of 10 foot boards is $124 \div 2 = 62$. Therefore, there were 28 8-foot boards.

7. (a) Answers will vary. One solution is

    $$\begin{array}{r} 379 \\ 462 \\ +158 \\ \hline 999 \end{array}$$

    (b) Yes. The digits in each column can be arranged in any order. (Other answers, such as 198 + 267 + 534, are also possible.)

    (c) No. The hundreds column digit must sum to 8 to allow for a carry from the tens column. If the digit 1 is not in the hundreds column, the smallest that this sum can be is 2 + 3 + 4 = 9. (Note that if 2, 3, and 4 are used, it will be impossible to construct a sum that does not require carrying from the tens digit. Thus, the digit 1 must be in the hundreds column.)

8. There are 9 ways to produce 21¢ in change. They are listed in order of number of pennies used:

    4N + P   2N + 11P
    D + 2N + P   D + 11P
    2D + P   N + 16P
    3N + 6P   21P
    D + N + 6P

9. $5 \times 4 \times 3 = 60$

10. The flower bed plus the walkway has a total area of $12 \times 14 = 168$ square feet, and the flower bed alone has an area of $8 \times 10 = 80$ square feet. The area of the walkway is $168 - 80 = 88$ square feet.

11. Karen's number is 9. Work backward. Add 7 to the result, 11, to obtain 18. Divide by 2 to obtain 9.

12. (a) Multiply by 5, then subtract 2.

    (b) Answers will vary. A good strategy is to give Jon consecutive whole numbers starting with 0.

13. (a) Multiply each term by 2 to get the next term: 3, 6, 12, 24, 48, 96.

    (b) Since $16 \div 4 = 2^2$, multiply each term by 2 to get the next term: 4, 8, 16, 32, 64, 128.

    (c) Since $216 \div 1 = 6^3$, multiply each term by 6 to get the next term: 1, 6, 36, 216, 1296, 7776.

    (d) Since $1250 \div 2 = 5^4$, multiply each term by 5 to get the next term: 2, 10, 50, 250, 1250, 6250.

    (e) Since $7 \div 7 = 1^5$, multiply each term by 1 to get the next term: 7, 7, 7, 7, 7, 7.

14. We make a table to show all possiblities and use the clues to delete those that are impossible and, hence, those that are certain. The steps in the argument are numbered and the numbers in the table indicate the corresponding conclusions at each step.

	Doctor	Engineer	Teacher	Lawyer	Writer	Painter
Kimberly	no (5)	no (7)	no (1)	yes (8)	no (1)	yes (3)
Terry	yes (5)	yes (7)	no (6)	no (8)	no (2)	no (3)
Otis	no (4)	no (6)	yes (6)	no (6)	yes (2)	no (3)

(1) By (b), Kimberly is neither the teacher nor the writer.

(2) By (e), Terry is not the writer. Therefore, Otis is the writer.

(3) By (f), neither Otis nor Terry is the painter. Therefore, Kimberly is the painter.

(4) By (g), Otis is not the doctor.

(5) By (d), since the doctor hired the painter (Kimberly) and the doctor is not Otis, Terry is the doctor and Kimberly is not the doctor.

(6) Since Kimberly is not the teacher and, by (a), the doctor (Terry) had lunch with the teacher, Otis is the teacher and Terry is not. Also, since Otis has just two jobs, it follows that he is neither the engineer nor the lawyer.

(7) By (c), the painter (Kimberly) is related to the engineer. Therefore, Kimberly is not the engineer and so Terry is.

(8) Since Terry is the doctor and engineer he is not the lawyer. Thus, finally, Kimberly is the lawyer and the table now shows who holds what jobs.

15. (a) In the nth equation, we add the "next" n odd numbers:
$21 + 23 + 25 + 27 + 29 = 125 = 5^3$
$31 + 33 + 35 + 37 + 39 + 41 = 216 = 6^3$
$43 + 45 + 47 + 49 + 51 + 53 + 55 = 343 = 7^3$

(b) 1, 3, 7, 13, 21, 31, 43, 57, 73, 91
Note that 3 is 2 larger than 1, 7 is 4 larger than 3, 13 is 6 larger than 7, and so on.

(c) $91 + 93 + 95 + 97 + 99 + 101 + 103 + 105 + 107 + 109 = 10^3$

16. (a) In the nth equation, we add the "next" n even numbers:
$14 + 16 + 18 + 20 = 4^3 + 4$
$22 + 24 + 26 + 28 + 30 = 5^3 + 5$
$32 + 34 + 36 + 38 + 40 + 42 = 6^3 + 6$

(b) $92 + 94 + 96 + 98 + 100 + 102 + 104 + 106 + 108 + 110 = 10^3 + 10$

17. (a) Each term is 3 more than its predecessor. Since 79 is 72 more than 7, and $72 \div 3 = 24$, there are 24 terms after the first term—for a total of 25 terms.

(b) Use Gauss's method:
$$\begin{aligned} s &= 7 + 10 + 13 + \cdots + 79 \\ s &= \underline{79 + 76 + 73 + \cdots + 7} \\ 2s &= 86 + 86 + 86 + \cdots + 86 \\ s &= \frac{25 \cdot 86}{2} = 1075 \end{aligned}$$

18. (a) 11th term: 3, 6, 12, 24, 48, 96, 192, 384, 768, 1536, 3072
This can also be determined by observing that $3072 = 2^{10} \cdot 3$.

(b) By adding the terms listed above, $S = 6141$.

(c) Duly observed.

(d) $$\begin{aligned} 2S &= 6 + 12 + 24 + \cdots + 3072 + 6144 \\ -S &= \underline{-3 - 6 - 12 - 24 - \cdots - 3072} \\ S &= -3 + 0 + 0 + 0 + \cdots + 0 + 6144 \\ S &= -3 + 6144 = 6141 \end{aligned}$$

19. $5 + 15 + 45 + 135 + 405 + 1215 + 3645 + 10{,}935 + 32{,}805 + 98{,}415 + 295{,}245 = 442{,}865$.
Alternately, use the method of problem 18(d):

$$3S = 15 + 45 + \cdots + 295{,}245 + 885{,}735$$
$$-S = -5 - 15 - 45 - \cdots - 295{,}245$$
$$\overline{}$$
$$2S = -5 + 0 + 0 + \cdots + 0 + 885{,}735$$
$2S = -5 + 885{,}735 = 885{,}730$
$S = 885{,}730 \div 2 = 442{,}865$

20. Complete the chart below and then generalize from the results.

Number of chords	Number of regions	Number of intersections	Number of segments
0	1 = 0 + 1	0	0
1	2 = 1 + 1	0	1
2	4 = 3 + 1	1	4
3	7 = 6 + 1	3	9
4	11 = 10 + 1	6	16
5	16 = 15 + 1	10	25
6	22 = 21 + 1	15	36
$\vdots$	$\vdots$	$\vdots$	$\vdots$
n	$\dfrac{n(n+1)}{2} + 1$	$\dfrac{n(n-1)}{2}$	n^2

(a) $\dfrac{n(n+1)}{2} + 1$ or $t_n + 1$

(b) $\dfrac{n(n-1)}{2}$ or t_{n-1}

(c) Each chord is divided into n segments, for a total of n^2 small segments.

21. In every case tried, the product of the "squared" entries is equal to the product of the circled entries. We guess that this is always true.

22. (a) $1 + 1 \cdot 2 = 3$

(b) $1 + 2 \cdot 2 + 1 \cdot 2^2 = 9$

(c) $1 + 3 \cdot 2 + 3 \cdot 2^2 + 1 \cdot 2^3 = 27$

(d) The left side of the nth equation is obtained by multiplying the entries in the nth row of Pascal's triangle by successive powers of 2, and the right side is 3^n. This suggests that
$P_0 + P_1 \cdot 2^1 + P_2 \cdot 2^2 + \cdots + P_n \cdot 2^n = 3^n$, where P_k is the kth element in the nth row of Pascal's triangle.

(e) $1 + 1 \cdot 3 = 4$
$1 + 2 \cdot 3 + 1 \cdot 3^2 = 16$
$1 + 3 \cdot 3 + 3 \cdot 3^2 + 1 \cdot 3^3 = 64$

(f) Part (e) suggests that $P_0 + P_1 \cdot 3^1 + P_2 \cdot 3^2 + \cdots + P_n \cdot 3^n = 4^n$.
Taken together, the results suggest that $P_0 + P_1 \cdot r^1 + P_2 \cdot r^2 + \cdots + P_n \cdot r^n = (r+1)^n$.
To check further, one might try another example: $1 + 4 \cdot 5^1 + 6 \cdot 5^2 + 4 \cdot 5^3 + 1 \cdot 5^4 = 1296 = 6^4$.
The result is true in general, but one cannot be sure of this without a mathematical proof.

23. (a) There are four "pigeonholes" (suits), so draw 5 cards.

    (b) If only 8 cards are drawn, there could be 2 of each suit. Therefore, draw 9 cards.

    (c) If one drew 48 cards, one might get everything except the aces. Therefore, to be absolutely sure of getting two aces, one must draw 50 cards.

24. 17, since it is possible that the first 16 books chosen are 4 each from the 4 types of books.

25. (a) Work backward. The minimum cost is $2500.

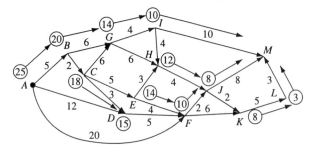

    (b) A B G I M or A B C D F J M

26. Awarding each item to the highest bidder, Judy gets the house, Joshua gets the motor home and painting, and JoAnn gets the automobile. Since these items have a total value of $1,309,000 and the fair shares total $1,127,000, each person's fair share is increased by $\frac{1}{4}(1,309,000 - 1,127,000) = 45,500$. Therefore Joshua should give $94,500 to Judy, $308,000 to John, and $267,500 to JoAnn.

	Judy	John	Joshua	JoAnn
Motor home	27,000	32,000	**35,000**	30,000
Automobile	19,000	18,000	23,000	**24,000**
House	**250,000**	200,000	220,000	230,000
Painting	900,000	800,000	**1,000,000**	700,000
Total	1,196,000	1,050,000	1,278,000	984,000
Fair Share	299,000	262,500	319,500	246,000
Fair Share + $45,500	344,500	308,000	365,000	291,500

27. (a)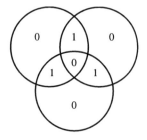

    1 1 0 1 0 0 0

(b)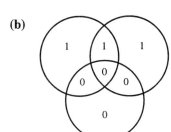

1 0 0 0 1 0 1

(c)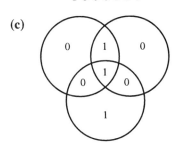

1 0 1 0 0 1 0

28. (a) Given the code word 1 1 1 1 0 0 0 1, we drew the associated Hamming diagram as shown.

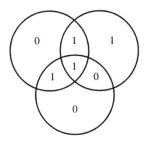

This reveals an error since two of the circles contain an odd number of 1s. To correct the error, we change the 1 common to these two circles but not to the third to a 0. This gives the diagram shown below and hence the correct code word 0 1 1 0 0 0 1.

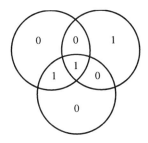

(b) The code word 0 1 0 1 1 0 1 corresponds to the Hamming diagram:

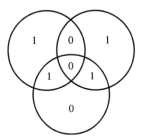

Since each circle contains an even number of 1s the code word is correct

(c) the code word 1 0 0 0 0 1 0 corresponds to the Hamming diagram

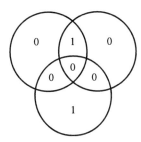

with an odd number of 1s in each circle. To correct it we change the 0 common to all three circles to a 1. This gives the diagram

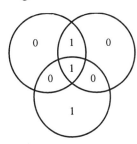

and the corrected code word 1 0 1 0 0 1 0.

Chapter 1 Test (page 75)

1. Note the pattern: As the "missing digit" on the left side of the equation moves to the right, the 2 on the right side of the equation moves to the right.

   $2,345,679 \times 9 = 21,111,111$
   $1,345,679 \times 9 = 12,111,111$
   $1,245,679 \times 9 = 11,211,111$
   $1,235,679 \times 9 = 11,121,111$
   $1,234,679 \times 9 = 11,112,111$
   $1,234,579 \times 9 = 11,111,211$
   $1,234,569 \times 9 = 11,111,121$
   $1,234,568 \times 9 = 11,111,112$

2. (a) Note first that $1 = 1^2$, $4 = 2^2$, $9 = 3^2$, ...; $0 = 0^3$, $1 = 1^3$, $8 = 2^3$,...; $0 = 0^2$, $1 = 1^2$, $9 = 3^2$, $36 = 6^2$, ...; and that 0, 1, 3, 6, ... are the triangular numbers. Moreover, the fourth equation, for example, can be written as $(3^2 + 1) + (3^2 + 2) + ... + 4^2 = 3^3 + 4^3 = 10^2 - 3^2 = t_4^2 - t_2^2$. Since this pattern holds for all four equations, we guess that the fifth equation should be
   $(4^2 + 1) + (4^2 + 2) + ... + 5^2 = 4^3 + 5^3 = 15^2 + 6^2 = t_5^2 - t_3^2$ or
   $17 + 18 + ... + 25 = 64 + 125 = 225 - 36$.
   Similarly, the sixth equation should be $(5^2 + 1) + (5^2 + 2) + ... + 6^2 = 5^3 + 6^3 = t_6^2 - t_4^2$
   or $26 + 27 + ... + 36 = 125 + 216 = 441 - 100$.
   Note that these are easily checked with your calculator.

   (b) Continuing the same pattern, the tenth equation should be
   $(9^2 + 1) + (9^2 + 2) + ... + 10^2 = 9^3 + 10^3 = t_{10}^2 - t_8^2$.
   Since $t_{10} = \frac{10 \cdot 11}{2} = 55$ and $t_8 = \frac{8 \cdot 9}{2} = 36$, this can be rewritten as
   $82 + 83 + ... + 100 = 729 + 1000 + 3025 - 1296$.

   (c) In general, the nth equation should be
   $[(n-1)^2 + 1] + [(n-1)^2 + 2] + ... + n^2 = (n-1)^3 + n^3$
   $= \left[\frac{n(n+1)}{2}\right]^2 - \left[\frac{(n-2)(n-1)}{2}\right]^2$ since $t_n = \frac{n(n+1)}{2}$ and $t_{n-2} = \frac{(n-2)(n-1)}{2}$.

3. Work backwards. Before meeting the third guard, he had $2 \cdot (1 + 2) = 6$ apples. Before meeting the second guard, he had $2 \cdot (6 + 2) = 16$ apples. Before meeting the first guard, he had $2 \cdot (16 + 2) = 36$ apples. He originally stole 36 apples.

4. The pigeons are most evenly distributed if one hole has 2 pigeons and one has 3 pigeons. This minimizes the number of pigeons in the hole with the most pigeons. The answer is 3 pigeons.

5. (a) Answers will vary. Each sum must be $\frac{2+7+12+27+32+37+52+57+62}{3} = 96$.
   One possibility is shown.

57	2	37
12	32	52
27	62	7

   (b) Answers will vary. A magic subtraction square can be obtained by interchanging the numbers at the ends of each diagonal in (a). One possibility is shown.

7	2	27
12	32	52
37	62	57

6. 10 days. (He reaches a *maximum* height of 3 feet on the first day, 4 feet on the second day, and so on.)

7. **(a)** $2 + 5 + 8 + 11 + 1 + 8 + 5 + 2 = 41 = 3^2 + 2 \cdot 4^2$
$2 + 5 + 8 + 11 + 14 + 11 + 8 + 5 + 2 = 66 = 4^2 + 2 \cdot 5^2$

(b) The middle number on the left side is $2 + 9 \cdot 3 = 29$.
$2 + 5 + 8 + \cdots + 26 + 29 + 26 + \cdots + 8 + 5 + 2 = 281 = 9^2 + 2 \cdot 10^2$

8. We note that 1, 3, 6, 10, 15, … are the triangular numbers. However, since we are only asked about the results, we concentrate on the numbers on the right sides of the equations. Consider the following table.

Number of equation	Result on right side
1	$1 = 1^2 = \left[\frac{(1+1)}{2}\right]^2$
2	$-2 = -1 \cdot 2 = -\left(\frac{2}{2}\right)\left(\frac{2}{2}+1\right)$
3	$4 = 2^2 = \left[\frac{(3+1)}{2}\right]^2$
4	$-6 = -2 \cdot 3 = -\left(\frac{4}{2}\right)\left(\frac{4}{2}+1\right)$
5	$9 = 3^2 = \left[\frac{(5+1)}{2}\right]^2$
6	$-12 = -3 \cdot 4 = -\left(\frac{6}{2}\right)\left(\frac{6}{2}+1\right)$

(a) Continuing the pattern, we guess that $s_{20} = -\left(\frac{20}{2}\right)\left(\frac{20}{2}+1\right) = -10 \cdot 11 = -110$.
Also we guess that $s_{21} = \left[\frac{(21+1)}{2}\right]^2 = 11^2 = 121$

(b) In general, we guess that $s_n = -\left(\frac{n}{2}\right) \cdot \left(\frac{n}{2}+1\right)$ for n even and $s_n = \left[\frac{(n+1)}{2}\right]^2$ for n odd.

9. **(a)**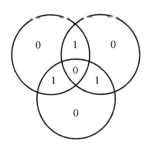

1 1 0 1 0 0 0

(b) Given:

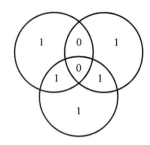

0 1 0 1 1 1 1

Corrected:

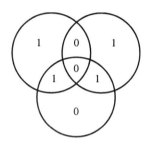

0 1 0 1 1 0 1

10. Work backwards. The lowest cost is 20, by following the rule EENNEEN.

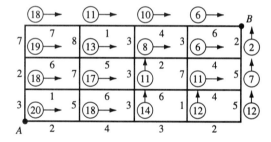

Chapter 2

Problem Set 2.1 (page 91)

1. (a) {Arizona, California, Idaho, Oregon, Utah}

2. (a) {1, i, s, t, h, e, m, n, a, o, y, c}

3. (a) {8, 9, 10, 11, 12}

   (c) {3, 6, 9, 12, 15, 18}

4. Answers will vary.

   (a) $\{x \in U | 11 \leq x \leq 14\}$ or
   $\{x \in U | 10 < x < 15\}$

   (c) $\{x \in U | x = 4n \text{ and } n \in N\}$

5. Answers may vary.

   (a) $\{x \in N | x \text{ is even and } x > 12\}$ or $\{x \in N | x = 2n \text{ for } n \in N \text{ and } n > 6\}$

6. (a) No, it is not clear which cities with these names are meant. For example, Moscow could be a city in Russia or in Idaho.

   (b) Yes, it is clear which four cities are included in the set.

7. (a) True. Both sets contain only the elements {t, w, e, n, t, y, o}.

   (c) True. The sets contain the same elements.

   (e) True. Every element in {1, 2, 3} is in {3, 2, 1}.

8.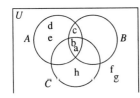

   (a) $B \cup C = \{a, b, c, h\}$

   (c) $A - B = \{d, e\}$

   (e) $\overline{A} = \{f, g, h\}$

9. (a) $M = \{45, 90, 135, 180, 225, 270, 315, ...\}$

   (b) $L \cap M = \{90, 180, 270, ...\}$
   This can be described as the set of natural numbers that are divisible by both 6 and 45, or as the set of natural numbers that are divisible by 90.

   (c) 90

11. (a) $A \cap B \cap C$ is the set of elements common to all three sets.

    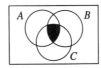

    (c) $(A \cap B) \cup C$ is the set of elements in C or in both A and B.

    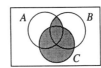

    (e) $A \cup B \cup C$ is the set of all elements in A, B, and/or C.

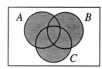

12. **(a)** *A* and *B* must be disjoint sets contained in *C*.

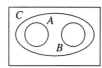

13. No. The statement $A \cup B = A \cup C$ implies only that the portions of *B* and *C* that are outside *A* must be the same or $B \subseteq A$ and $C \subseteq A$). For example, *A* and *B* could be disjoint sets with $C = A \cup B$, as when $A = \{1\}, B = \{2\}$ and $C = \{1, 2\}$.

15. **(a)** Since $A \cap B = \{6, 12, 18\}$,
$\overline{A \cap B} = \{1, 2, 3, 4, 5, 7, 8, 9, 10, 11, 13, 14, 15, 16, 17, 19, 20\}$.
Since $\overline{A} = \{1, 3, 5, 7, 9, 11, 13, 15, 17, 19\}$ and
$\overline{B} = \{1, 2, 4, 5, 7, 8, 10, 11, 13, 14, 16, 17, 19, 20\}$,
$\overline{A} \cup \overline{B} = \{1, 2, 3, 4, 5, 7, 8, 9, 10, 11, 13, 14, 15, 16, 17, 19, 20\}$
Since $A \cup B = \{2, 3, 4, 6, 8, 9, 10, 12, 14, 15, 16, 18, 20\}$, $\overline{A \cup B} = \{1, 5, 7, 11, 13, 17, 19\}$.
Since $\overline{A} = \{1, 3, 5, 7, 9, 11, 13, 15, 17, 19\}$ and
$\overline{B} = \{1, 2, 4, 5, 7, 8, 10, 11, 13, 14, 16, 17, 19, 20\}$,
$\overline{A} \cap \overline{B} = \{1, 5, 7, 11, 13, 17, 19\}$.

(b) Since $\overline{A \cap B} = \overline{A} \cup \overline{B}$ and $\overline{A \cup B} = \overline{A} \cap \overline{B}$ from part (a), DeMorgan's Laws hold for these sets.

16. **(a)** $\{(1, a), (1, b), (2, a), (2, b)\}$

(c) $\{(6, t)\}$

17. **(a)** $\{d, e\} \times \{4, 6\}$

(c) $N \times N$

(e) $\emptyset \times \emptyset$, $\{2, 9\} \times \emptyset$, $\emptyset \times \{a, b, c\}$

19. **(a)** Although the elements of a set can be written in any order, the entries in an ordered pair must be written in the correct order (hence, the term *ordered* pair). The first component in an ordered pair is an element of the first set and the second component is an element of the second set. For example, if
$A = \{1, 2\}$ and $B = \{a, b\}$, then
$A \times B = \{(1, a), (1, b), (2, a), (2, b)\}$ but
$B \times A = \{(a, 1), (a, 2), (b, 1), (b, 2)\}$.

(b) $C \times D = D \times C$ if and only if $C = D$ or at least one of the sets is $\emptyset$.

20. **(a)** $R \cap C$ is the set of red circles. Draw two red circles—one large and one small.

(c) $T \cup H$ is the set of shapes that are either triangles or hexagons. Draw two large triangles, two large hexagons, two small triangles, and two small hexagons—a red and a blue of each.

(e) $B - C$ is the set of blue shapes that are not circles. Draw two blue hexagons and two blue triangles—a large and a small of each.

21. **(a)** $L \cap T$

(c) $S \cup T$

22. (a) 8 subsets: $\emptyset$, {a}, {b}, {c}, {a, b}, {a, c}, {b, c}, {a, b, c}

(c) 8 subsets—same as the subsets of {a, b, c}.

(e) $2^{26} = 67,108,864$

23. (a) 8 regions

(c) Verify that $A \cap \overline{B} \cap \overline{C} \cap D$ has no region by observing that the region for $A \cap D$ is entirely contained in $B \cup C$. Likewise, the region for $B \cap C$ is entirely contained in $A \cup D$. The other missing region is $\overline{A} \cap B \cap C \cap \overline{D}$.

24. Use the notation suggested in the hint. Write each statement in terms of set operations and relations.

Statement Number	Set Statement
1	$N \subseteq S$
2	$A \subseteq M$
3	$S \cap L = \emptyset$
4	$M \subseteq N$

From statements 1, 2, and 4, we have $A \subseteq M \subseteq N \subseteq S$. Since S and L do not have any members in common, and A is a subset of S, this implies that A and L do not have any members in common. (In other words, all birds in this aviary are ostriches, which do not live on mince pie.) Therefore, no birds in this aviary live on mince pie.

26. (a) 15 ways, as follows:

Configuration	Partitions
One 4-element set (D itself) (1 partition)	{a, b, c, d}
One 3-element set, one 1-element set (4 partitions)	{a, b, c} $\cup$ {d}
	{a, b, d} $\cup$ {c}
	{a, c, d} $\cup$ {b}
	{b, c, d} $\cup$ {a}
Two 2-element sets (3 partitions)	{a, b} $\cup$ {c, d}
	{a, c} $\cup$ {b, d}
	{a, d} $\cup$ {b, c}
One 2-element set, two 1-element sets (6 partitions)	{a, b} $\cup$ {c} $\cup$ {d}
	{a, c} $\cup$ {b} $\cup$ {d}
	{a, d} $\cup$ {b} $\cup$ {c}
	{b, c} $\cup$ {a} $\cup$ {d}
	{b, d} $\cup$ {a} $\cup$ {c}
	{c, d} $\cup$ {a} $\cup$ {b}
Four 1-element sets (1 partition)	{a} $\cup$ {b} $\cup$ {c} $\cup$ {d}

27. (a) Since each loop corresponds to C, H, T, S, L, R, or B, we can use an elimination process to show that there is only one possibility. The set on the left cannot be a color (because it would be blue—but there are blue figures outside also) or a size (because it contains large and small figures), so it must be a shape—circles. The set on the right contains more than one size and more than one shape, so it must be a color—blue.

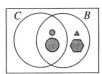

28. (a) There are 6 choices for shape, 2 choices for size, and 3 choices for color, so the number of pieces is $6 \times 2 \times 3 = 36$.

    (b) $6 \times 2 \times 3 \times 2 = 72$

30. Note in this diagram, O refers to the region outside both circles.

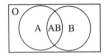

32. Answers will vary. Consult a thesaurus.

34. Polya's Model
    Understand the Problem
    A perfect square is a number which can be obtained by multiplying an integer by itself. For example, $25 = 5 \times 5$, $64 = 8 \times 8$, etc. The problem is to create a list using the integers from 1 to 15 inclusive so that any two adjacent numbers have a sum which is a perfect square.
    Devise Strategies
    Many of the numbers can be added to more than one number to produce a square. For example, 3 can be added to 6 or 13, but the numbers 8 and 9 each have only one number to which they can be added to produce a square. Therefore, 8 and 9 must go at the "ends" of the list.
    Carry Out the Plan
    If 8 is placed at the front of the list, it must be followed by 1, etc.: 8, 1, 15, 10, 6, 3, 13, 12, 4, 5, 11, 14, 2, 7, 9.
    Look Back
    A key to solving the problem is the recognition that only two numbers 8 and 9 each have only one number from 1 to 15 inclusive to which they can be added to produce a perfect square.

JUST FOR FUN   Disk Discoveries (page 99)

The discs differ only in color and size, so we compare these attributes:

Disc	Color	Size
Large red	Does not match	Matches 1 other disc
Small blue	Matches 1 other disc	Does not match
Large blue	Matches 1 other disc	Matches 1 other disc

Clearly, the large blue disc is the most like each of the other discs. If these were part of an original four disc set, the fourth disc was probably a red disc the same size as the small blue disc.

Problem Set 2.2 (page 103)

1. (a) 13: ordinal
       first: ordinal

2. Note that $B = \{$Huey, Dewey, Louie$\}$. Answers will vary. Two possibilities are shown.

   h ↔ Huey   d ↔ Dewey   l ↔ Louie
   h ↔ Louie   d ↔ Huey   l ↔ Dewey

3. The cardinal number of seats in the auditorium is less than the cardinal number of the potential number of audience members. In other words, the concert organizers want to make sure that there are enough seats.

4. Answers will vary. One possibility is shown.

   (a)   a   b   c
         ↕   ↕   ↕
         □   △   ★

5. (a) $n(A) = 15$ because $A = \{20, 21, 22, 23, 24, 25, 26, 27, 28, 29, 30, 31, 32, 33, 34\}$.

   (c) $n(C) = 2$ because $C = \{3, 8\}$.

7. Yes, $T = \{3, 6, 9, ...\}$, $F = \{5, 10, 15, ...\}$.
   $3 \leftrightarrow 5, 6 \leftrightarrow 10, 9 \leftrightarrow 15, ... 3k \leftrightarrow 5k, ...$

8. (a) Finite. The number of grains of sand is large, but finite.

9. (a) The set $S = \{n \mid n = m^3 \text{ for some } m \in N\}$
   $= \{1, 8, 27, 64, ..., n^3, ...\}$ is infinite because it can be put in one-to-one correspondence with one of its proper subsets:

   S: 1   8   27   64   ... n^3
               ↕   ↕   ↕   ↕
   Subset: 1 8^3 27^3 64^3 ... $(n^3)^3$

   This method is not discussed in the text.

10. (a) Answers may vary. For example, $Q_1 \leftrightarrow Q_2$, $Q_3 \leftrightarrow Q_4$, and so on. (Note that P need not be the center—it can be any fixed point inside the small circle.)

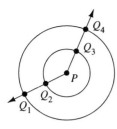

(c) Answers may vary. For example, $Q_1 \leftrightarrow Q_2$, etc.

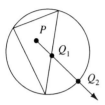

11. (a) True. A set B cannot have fewer elements than its subset A.

(c) True. Since A and B are finite, the statement $n(A - B) = n(A)$ means that removing any elements of B from A has no effect—that is, $A \cap B = \varnothing$.

13. Since $n(A \times B) = n(A) \times n(B)$, the requirement is that $n(A) \times n(B) = 12$. The possibilities are:

$n(A)$	$n(B)$
1	12
2	6
3	4
4	3
6	2
12	1

14. (a) $n(A \cap B) \leq n(A)$. The set $A \cap B$ contains only the elements of A that are also elements of B. That is, $A \cap B \subseteq A$. Thus, $A \cap B$ cannot have more elements than A.

15. (a) For each ordered triple, there are 4 choices for the first component, 2 choices for the second component, and 5 choices for the third component. Therefore, the total number of ordered triples is $4 \times 2 \times 5 = 40$.

16. Use the fact that the eight regions pictured below are mutually disjoint sets and apply the strategy of working backwards, that is, start with $n(A \cap B \cap C) = 7$. Next use $n(A \cap B)$, $n(B \cap C)$, and $n(A \cap C)$ to find the values 10, 5, and 8. Then use $n(A)$, $n(B)$, and $n(C)$ to find the values 15, 28, and 10, and finally $n(U)$ to find the value 17.

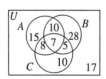

18. (a) 1, 4, 9, 16, 25

19. (a)   37   61

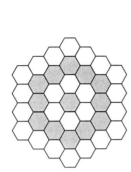
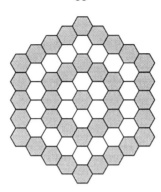

20. (a) No. Since he lives forever, Joe will eventually earn enough to pay all his bills through any given week even though he falls further and further behind.

(b) Joe pays 3 weeks of bills every 10 weeks. Since weeks 100 to 102 make up the 34th group of 3 weeks, and $34 \times 10 = 340$, he will pay these bills after week 340.

22. (a) If $A = \{a_1, a_2, a_3, ...\}$, then the correspondence:

$$\begin{array}{cccc} a_1 & a_2 & a_3 & ... & a_k & ... \\ \updownarrow & \updownarrow & \updownarrow & & \updownarrow & \\ a_1 & a_2 & a_3 & ... & a_k & ... \end{array}$$

shows that $A \sim A$.

(b) $A \sim B$, then there must be a 1-1 correspondence:

$$\begin{array}{ccc} a_1 & a_2 & ... & a_n \\ \updownarrow & \updownarrow & & \updownarrow \\ b_1 & b_2 & ... & b_n \end{array}$$

To show this implies $B \sim A$, just reverse the correspondence:

$$\begin{array}{ccc} b_1 & b_2 & ... & b_n \\ \updownarrow & \updownarrow & & \updownarrow \\ a_1 & a_2 & ... & a_n \end{array}$$

(c) If $A \sim B$ and $B \sim C$, then the following 1-1 correspondences must exist:

$$\begin{array}{ccc} a_1 & a_2 & ... & a_n \\ \updownarrow & \updownarrow & & \updownarrow \\ b_1 & b_2 & ... & b_n \end{array}$$

$$\begin{array}{ccc} b_1 & b_2 & ... & b_n \\ \updownarrow & \updownarrow & & \updownarrow \\ c_1 & c_2 & ... & c_n \end{array}$$

By combining the correspondences, we obtain:

$$\begin{array}{ccc} a_1 & a_2 & ... & a_n \\ \updownarrow & \updownarrow & & \updownarrow \\ c_1 & c_2 & ... & c_n \end{array}$$

Thus, $A \sim C$ and the equivalence relation is transitive.

24. (a) The diagram shown below gives $36 + x = 40$, so $x = 4$. (This result can also be obtained using guess and check.) 4 students have been to all three countries.

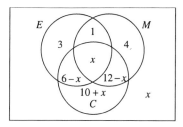

(b) $10 + 4 = 14$ students

26. (a) Making an orderly list, we obtain the following:
    a ↔ 1, b ↔ 2, c ↔ 3
    a ↔ 1, b ↔ 3, c ↔ 2
    a ↔ 2, b ↔ 1, c ↔ 3
    a ↔ 2, b ↔ 3, c ↔ 1
    a ↔ 3, b ↔ 1, c ↔ 2
    a ↔ 3, b ↔ 2, c ↔ 1

(b) Taking the elements in the first set in the order in which they appear, p can be matched with any one of four choices, namely, 1, 2, 3, 4; q can be matched with any of the remaining three choices, r with either of the remaining two choices, and s with the last remaining choice. Therefore, the number of different 1-1 correspondences is $4 \times 3 \times 2 \times 1 = 24$

27. (a)

r	Subsets with r elements	Number of subsets with r elements
0	∅	1
1	{a}, {b}, {c}, {d}	4
2	{a, b}, {a, c}, {a, d}, {b, c}, {b, d}, {c, d}	6
3	{a, b, c}, {a, b, d}, {a, c, d}, {b, c, d}	4
4	{a, b, c, d}	1
5	None	0

29. Use the Venn diagram below to solve the problem.

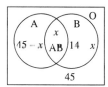

$104 - x = 100$
$x = 4$
Four people are of type AB. The completed Venn diagram is shown below.

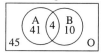

33. $U = \{0, 1, 2, 3, ..., 25\}$, $A = \{0, 2, 4, ..., 24\}$,
$B = \{0, 1, 4, 9, 16, 25\}$,
$C = \{1, 2, 3, 4, 6, 8, 12, 16, 24\}$

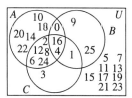

34. (a) $\overline{A} = \{b, c, f, g\}$

(c) $A \cup \overline{B} = \{a, d\}$

(e) $\overline{A} \cap \overline{B} = \{b\}$

(g) {(a, c), (a, e), (a, f), (a, g), (d, c), (d, e), (d, f), (d, g), (e, c), (e, e), (e, f), (e, g)}

35. (a)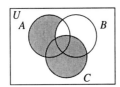

JUST FOR FUN Paper Clip Comparison
(page 113)

Many people will say just one more and will be surprised when shown that the answer is two more.

Problem Set 2.3 (page 116)

1. (a) (i) $A \cup B$ = {apple, berry, peach, lemon, lime} so $n(A \cup B) = 5$.

    (ii) $A \cup C$ = {apple, berry, peach, lemon, prune} so $n(A \cup C) = 5$.

    (iii) $B \cup C$ = {lemon, lime, berry, prune} so $n(B \cup C) = 4$ since lemon is a member of both sets.

    (b) (ii) and (iii) because the two sets in each case are not disjoint, that is, they have at least one member in common.

3. (a) B may contain 4, 5, 6, 7, or 8 elements. If B were to contain more than 8 elements, then $n(A \cup B)$ would be greater than 8 which contradicts the fact that $n(A \cup B) = 8$. Similarly, if B were to contain less than 4 elements, $n(A \cup B)$ would be less than 8.

    (b) If $(A \cap B) = \emptyset$, then $n(A) + n(B) = n(A \cup B)$, so $n(B) = 4$.

4. (a)

7. (a) Closed (since the sum of two positive multiples of 5 is a larger multiple of 5)

    (c) Closed (since $0 + 0 = 0$)

    (e) Closed (since the sum of two numbers that are ≥ 19 is a larger number, which must be ≥ 19)

9. This can be justified by repeated application of the associative and commutative properties.

11. (a) $5 + 7 = 12$ $12 - 7 = 5$
        $7 + 5 = 12$ $12 - 5 = 7$

    (b) $4 + 8 = 12$ $12 - 8 = 4$
        $8 + 4 = 12$ $12 - 4 = 8$

12. (a) $419 + x = 627$ or $x + 419 = 627$

13. (a) 9

    (c) 0

    (e) 0, 1, 2, 3, 4, 5

    (g) 8, 9, 10, 11, ...

15. (a) Comparison

    (b) Missing addend

    (c) Take-away

    (d) Comparison

17. Jeff has read through page 240. Therefore, the number of pages is $257 - 240 = 17$ pages or $257 - 241 + 1 = 17$ pages.

18. Use the guess and check method. Some answers will vary.

    (a) $(8 - 5) - (2 - 1) = 2$

    (c) $((8 - 5) - 2) - 1 = 0$

19. **(a)** First fill in the squares by noting that 3 − 1 = 2, 4 − 2 = 2, and 7 − 2 = 5. Then complete the circles.

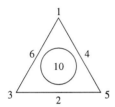

20. Write all possible combinations of 3 different numbers whose sum is the indicated number in the triangle. Then, place numbers which occur in more than one sum at the vertices as shown below.

(a) 1 + 3 + 6
2 + 3 + 5
4 + 5 + 1

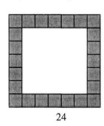

22.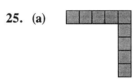

$n(A \cup B) = n(A) + n(B) - n(A \cap B)$

Note that elements of $A \cap B$ are counted twice when calculating $n(A) + n(B)$. We compensate by subtracting $n(A \cap B)$, giving $n(A \cup B)$.

24. **(a)** There will be 7 rows corresponding to the numbers 0, 1, 2, 3, 4, 5, 6. The number of dominoes is
1 + 2 + 3 + 4 + 5 + 6 + 7 = 28.

25. **(a)**

9

(c) $100^2 = 10,000$

26. **(a)** $d_4 = 24$, $d_5 = 32$

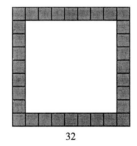

27. **(a)** $5^3 - 3^3 = 98$

28. {0}, since 0 − 0 = 0.

30. **(a)** Use the formula $t_n = \dfrac{n(n+1)}{2}$.

n	1	2	3	4	5	6	7	8	9	10	11	12	13	14	15
t_n	1	3	6	10	15	21	28	36	45	55	66	78	91	105	120

(b) 11 = 10 + 1
12 = 6 + 6
13 = 10 + 3
14 = 10 + 3 + 1
15 = 15
16 = 15 + 1
17 = 15 + 1 + 1
18 = 15 + 3
19 = 10 + 6 + 3
20 = 10 + 10
21 = 21
22 = 21 + 1
23 = 10 + 10 + 3
24 = 21 + 3
25 = 15 + 10

31. First find the top row and the left column. For example, the fourth entry in the left column must be 6 − 2 = 4, and then the third entry in the top row is 5 − 4 = 1, and so on. The completed table is shown.

+	5	4	1	6	9	2	0	8	7	3
3	8	7	4	9	12	5	3	11	10	6
9	14	13	10	15	18	11	9	17	16	12
6	11	10	7	12	15	8	6	14	13	9
4	9	8	5	10	13	6	4	12	11	7
0	5	4	1	6	9	2	0	8	7	3
7	12	11	8	13	16	9	7	15	14	10
5	10	9	6	11	14	7	5	13	12	8
2	7	6	3	8	11	4	2	10	9	5
1	6	5	2	7	10	3	1	9	8	4
8	13	12	9	14	17	10	8	16	15	11

34. (a) $n(A \cap B) = 0 + 2 = 2$

   (c) $n(\overline{A} \cap C) = 6 + 4 = 10$

   (e) $n(A - C) = 5 + 0 = 5$

36. 5, since $3 \times \square = 15$ means there are $\square = 5$ choices at Brownsport.

Problem Set 2.4 (page 133)

1. (a) $3 \times 5 = 15$

   (c) $4 \times 10 = 40$

   (e) $3 \times 4 = 12$

2. (a) Array model

   (c) Measurement model

4. (a) Each of the *a* lines from set *A* intersects each of the *b* lines from set *B*. Since the lines for *A* are parallel and the lines for *B* are parallel, the intersection points will all be distinct.

   (b)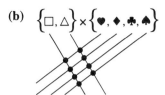

5. (a) Not closed. For example, $2 \times 2 = 4$, which is not in the set.

   (c) Not closed. For example, $2 \times 4 = 8$, which is not in the set.

   (e) Closed. The product of any two odd whole numbers is always another odd whole number.

   (g) Closed. $2^a \times 2^b = 2^{a+b}$ for any whole numbers *a* and *b*.

7. (a) Commutative property of multiplication

   (c) Multiplication by zero property

   (e) Associative property of multiplication

8. (a) Commutative property: $5 \times 3 = 3 \times 5$

9. (a)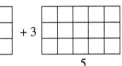

10.

a	a·d	a·e	a·f
b	b·d	b·e	b·f
c	c·d	c·e	c·f
	d	e	f

$(a+b+c) \cdot (d+e+f) = a \cdot d + a \cdot e + a \cdot f + b \cdot d + b \cdot e + b \cdot f + c \cdot d + c \cdot e + c \cdot f$

12. Each nut-and bolt pair cost 86¢ + 14¢ = $1.00. Therefore, 18 pairs cost $18.00.

13. (a) Distributive property of multiplication over addition: $7 \cdot 19 + 3 \cdot 19 = (7+3) \cdot 19 = 10 \cdot 19 = 190$

14. (a) $18 \div 6 = 3$

15. (a) $4 \times 8 = 32$, $8 \times 4 = 32$, $32 \div 8 = 4$, $32 \div 4 = 8$

16. (a) Repeated subtraction

17. (a) Since 78 - 13 = = = = = = gives 0, $78 \div 13 = 6$.

   (c) Since 96 -14 = = = = = = gives 12, $96 \div 14 = 6$ R 12.
   (*Note*: Stop pressing = when the result is less than the divisor, 14.)

18. Answers will vary. One counterexample is shown in each case.

  (a) Choose $a = 3$, $b = 2$: $3 \div 2$ is not a whole number.

19. (a) If $\frac{a}{b}$ and $\frac{d}{b}$ are defined, then there exist whole numbers r and s such that $\frac{a}{b} = r$ and $\frac{d}{b} = s$. But then $a = br$ and $d = bs$ so that $a + d = br + bs = b(r + s)$. This implies that $\frac{(a+d)}{b}$ is defined, and also that $\frac{a+d}{b} = r + s = \frac{a}{b} + \frac{d}{b}$.

20. (a) $y = (5 \cdot 5) + 4 = 25 + 4 = 29$

21. (a) $3^{20} \cdot 3^{15} = 3^{20+15} = 3^{35}$

  (c) $(3^2)^5 = 3^{2 \cdot 5} = 3^{10}$

  (e) $y^3 \cdot z^3 = (y \cdot z)^3$ or $(yz)^3$

22. (a) $8 = 2 \cdot 2 \cdot 2 = 2^3$

  (c) $1024 = 2^{10}$

23. (a) $4^2 = (2^2)^2 = 2^{2 \cdot 2} = 2^4$
  However, the power operation is *not* commutative because there are numbers for which $a^b \neq b^a$. For example, $2^3 \neq 3^2$ because $2^3 = 8$ and $3^2 = 9$.

24. Use the guess and check method.

  (a) $m = 4$

  (c) $p = 10$

25. The large square has side length $a + b$ so its area is $(a+b)^2$. But the square is divided into four regions with areas: a^2, ab, ab, and b^2, showing that the total area can also be expressed by $a^2 + 2ab + b^2$.

27. The key to understanding this problem is to notice that there are two overlapping a-by-a squares in the original figure, and the region of overlap has area $(a-b)^2$. Each expression is the area of the large square plus the area of the overlapping region, as shown below.

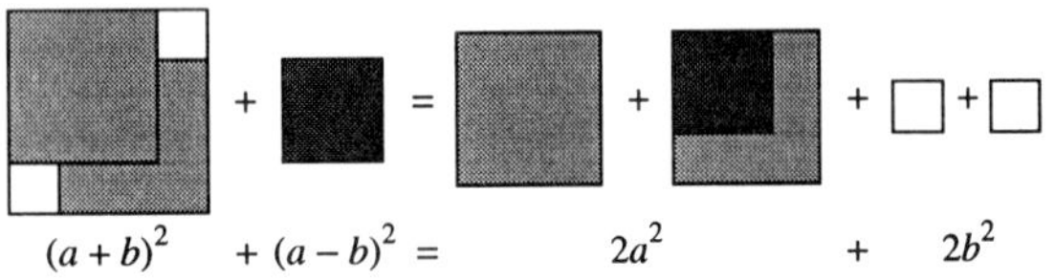

$(a+b)^2 + (a-b)^2 = 2a^2 + 2b^2$

29. (a) The quantities are equal because $4 \cdot (5 - 2) = 4 \cdot 3 = 12$ and $4 \cdot 5 - 4 \cdot 2 = 20 - 8 = 12$.

  (b) The diagram below makes it clear that $a(b - c) = ab - ac$.

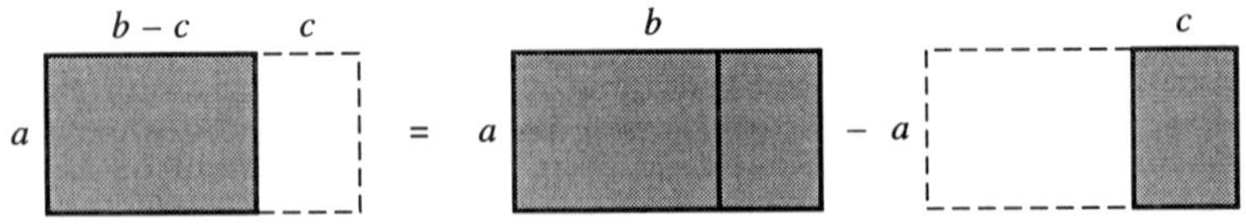

  Alternately, $a(b - c) + ac = a((b - c) + c) = ab$, so $a(b - c) = ab - ac$.

31. $2(1 + 2 + 3 + 4 + 5 + 6 + 7 + 8 + 9 + 10 + 11 + 12) = \frac{(2)(12 \times 13)}{2} = 156$ times

33.

n	1	2	3	4	5	6	7	8	9	10	11	12	13	14	15
(a) nth oblong number	2	6	12	20	30	42	56	72	90	110	132	156	182	210	240
(b) nth triangular	1	3	6	10	15	21	28	36	45	55	66	78	91	105	120

34. (a)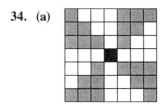

36. (b) $(10a+5)^2 = (10a+5)(10a+5)$
$= (10a+5) \cdot 10a + (10a+5) \cdot 5$
$= 100a^2 + 50a + 50a + 25$
$= 100a^2 + 100a + 25$
$= (a^2 + a)100 + 25$
$= a(a+1)100 + 25$

For any whole number a, $10a + 5$ is a whole number which ends in the digit 5 and conversely.

38. $2 + 7 = 9, 7 + 2 = 9, 9 - 7 = 2, 9 - 2 = 7$

40. $2 \times 3 = 6, \ 3 \times 2 = 6$

JUST FOR FUN   Red and Green Jelly Beans (page 141)

They are the same. After the first move, jar G contains 20 red jellybeans. If r is the number of red jellybeans that are then moved back to jar R, then $(20 - r)$ red jellybeans are left in jar G, and $(20 - r)$ green jellybeans have been moved to jar R.

Problem Set 2.5 (page 146)

1. (a) It depends on your point of view regarding the question, "Is a person his or her own friend?" If the answer is always yes, the relation is reflexive. If the answer is no, the relation is not reflexive. Most dictionaries suggest that the answer should be no.

   (c) No. A person cannot be his own father.

   (e) No. Some people (including residents of Washington, DC, and Puerto Rico) do not live in any state.

3. (a) Not transitive. It is possible that Abe and Betty are friends, Betty and Charlie are friends, but Abe and Charlie are not friends.

   (b) Transitive. If Mary is Josie's sister and Josie is Ralph's sister, then Mary is Ralph's sister.

   (c) Not transitive. The father of a father is a grandfather.

   (d) Transitive. If Xavier is in the same grade as Yvette and Yvette is in the same grade as Zelda, then Xavier is in the same grade as Zelda.

   (e) Transitive. If Xavier lives in the same state as Yvette and Yvette lives in the same state as Zelda, then Xavier lives in the same state as Zelda.

4. (a) Reflexive. A number has the same parity as itself.

   (c) Not reflexive. 0 is not related to itself because $0 \cdot 0 \not> 0$.

5. (a) Symmetric. If x has the same parity as y, then y has the same parity as x.

   (b) Not symmetric. For example, $4 | 12$ but $12 \nmid 4$.

   (c) Symmetric. If $xy > 0$, then $yx > 0$.

7. (a) Not reflexive. Nobody wins in a tie.
   Not symmetric. Rock wins against scissors, but scissors do not win against rock.
   Not transitive: Scissors win against paper and paper wins against rock, but scissors do not win against rock.

8. (a) Reflexive only

9. Answers will vary.

   (a)

10. (a) No. The relation is not symmetric.

    (c) No. The relation is not symmetric.

11. (a) $g(0) = 5 - 2(0) + 0^2 = 5$
       $g(1) = 5 - 2(1) + 1^2 = 4$
       $g(2) = 5 - 2(2) + 2^2 = 5$
       $g(3) = 5 - 2(3) + 3^2 = 8$
       $g(4) = 5 - 2(4) + 4^2 = 13$

    (b) {4, 5, 8, 13}

13. (a) $A = 2x \cdot x$ or $A = 2x^2$

    (b) $P = 2x + 2(2x)$ or $P = 6x$

    (c) Since $D^2 = x^2 + (2x)^2 = 5x^2$, $D = \sqrt{5}\,x$.

15. (a) Note that each arrangement has an additional picket and an additional row of dots.

       5   6

    (b)
Arrangement	1	2	3	4	5	6	7	8
Number of dots	1	3	8	15	24	35	48	63

    (c) $d = f(n) = n^2 - 1$ (except for the case $n = 1; f(1) = 1$)

17. (a) $19 \div 3 = 6$ R 1 and $31 \div 3 = 10$ R 1, so $19 \equiv 31$.
       $28 \div 3 = 9$ R 1 but $15 \div 3 = 5$ R 0, so $28 \not\equiv 15$.

    (b) Every element of E_0 is a multiple of 3 and so has a remainder of 0 when divided by 3. Every whole number that has a remainder of 0 when divided by 3 is in E_0, so E_0 is an equivalence class of the relation $\equiv$.

    (c) Every element of E_1 has a remainder of 1 when divided by 3, and every whole number that has a remainder of 1 when divided by 3 is in E_1, so E_1 is an equivalence class of the relation $\equiv$.

    (d) $E_2 = \{n \mid n = 3k + 2 \text{ and } k \in W\} = \{2, 5, 8, 11, ...\}$

18. (b) Suppose $a, b, c, d, e, f \in W$. The reflexive property holds: $(a, b) \square (a, b)$ because $a + b = b + a$. The symmetric property holds: If $(a, b) \square (c, d)$, then $a + d = b + c$, so $c + b = d + a$. Then $(c, d) \square (a, b)$. The transitive property holds:
If $(a, b) \square (c, d)$ and $(c, d) \square (e, f)$ then

$$a + d = b + c$$
and
$$c + f = d + e.$$

Adding gives $\quad a + d + c + f = b + c + d + e.$

Subtracting c and d from both sides gives $\quad a + f = b + e.$

Therefore, $(a, b) \square (e, f)$.

Since the reflexive, symmetric, and transitive properties all hold, $\square$ is an equivalence relation.

(d) $(a, b) \square (1, 0)$ if, and only if, $a + 0 = b + 1$.

Therefore, $E_1 = \{(a, b) | a = b + 1, a \in W, b \in W\}$
$= \{(1, 0), (2, 1), (3, 2), (4, 3), ...\}$
$= \{(m, m-1) | m \in N\}.$

20. (a) $T_5 = 2 + 4 + 7 = 13$
$T_6 = 4 + 7 + 13 = 24$
$T_7 = 7 + 13 + 24 = 44$
$T_8 = 13 + 24 + 44 = 81$
$T_9 = 24 + 44 + 81 = 149$
$T_{10} = 44 + 81 + 149 = 274$

The completed table is shown.

n	1	2	3	4	5	6	7	8	9	10
T_n	1	2	4	7	13	24	44	81	149	274

21. (a) 4 trains of length 3: WWW, RW, WR, G
7 trains of length 4: WWWW, RWW, WRW, WWR, RR, GW, WG
13 trains of length 5: WWWWW, RWWW, WRWW, WWRW, WWWR, RRW, RWR, WRR, GWW, WGW, WWG, GR, RG

The completed table is shown.

length of train, n	1	2	3	4	5
Number of trains, T_n	1	2	4	7	13

(c) Use the results of part (a) or Problem 20. $T_6 = 24$, $T_7 = 44$, $T_8 = 81$

22. (a) Yes.
$S_5 = 1^2 + 2^2 + 3^2 + 4^2 + 5^2 = 55 = \dfrac{10 \cdot 11 \cdot 12}{24}$
$S_6 = 1^2 + 2^2 + 3^2 + 4^2 + 5^2 + 6^2 = 91 = \dfrac{12 \cdot 13 \cdot 14}{24}$

(b) $S_n = \dfrac{2n(2n+1)(2n+2)}{24} = \dfrac{n(2n+1)(n+1)}{6}$

23. (a)

Size of the array	Number of squares of size					Total number of squares
	1×1	2×2	3×3	4×4	5×5	
1×1	1					1
2×2	4	1				5
3×3	9	4	1			14
4×4	16	9	4	1		30
5×5	25	16	9	4	1	55

24. (c) Each cevian from vertex C crosses p cevians from vertex A, q cevians from vertex B, and the additional segment $\overline{AB}$. Since no three cevians intersect at the same point, $p + q + 1$ additional regions are created for each cevian from C. Thus the number of regions is
$(p + 1)(q + 1) + (p + q + 1)r$
$= pq + pr + qr + p + q + r + 1$.

25. $L = mw + b$
$10 = m \cdot 0 + b$, so $b = 10$.
$14 = m \cdot 2 + b = 2m + 10$, so $m = 2$.
The constants are $b = 10$ and $m = 2$.

27. 100 mi/hr = 528,000 ft/.hr = $146.\overline{6}$ ft/sec.
Since $146.\overline{6} = 32t$, $t = \frac{146.\overline{6}}{32} = 4.58\overline{3}$.
It takes $4.58\overline{3}$ seconds.

29. 760 mi/hr = $0.2\overline{1}$ mi/sec $\doteq \frac{1}{5}$ mi/sec.
Thus, $d = \frac{1}{5}t$.

30. (a) The passenger is charged for 5 miles:
$2.15 + 4($1.25) = 7.15.

31. (a) $g = 2w + 2h$

32. (a) The customer is charged for 2 ounces:
32¢ + 23¢ = 55¢.

33. (a) From the tax table, the tax is $7075.

(b) If his earnings had been $100 lower, his tax would have been $7047. Since $7075 - 7047 = 28$, we conclude that $28 of the last $100 were paid in income tax.

34. (a) Since $3217.50 + 0.28(35,241 - 21,450) = 7078.98$, his tax would be $7078.98.

36. There are 4 prizes, each of which goes into 2500 boxes. It is possible (but unlikely) that the first 7500 boxes you buy could contain only 3 kinds of prizes, 2500 of each. To be certain of having all 4 prizes, you should buy 7501 boxes.

38. (a) $n(A \cup B) = n(\{a, b, c, d, e, f\}) = 6$

(b) $n(B \cap C) = n(\{b, e\}) = 2$

(c) $n(\overline{B}) = n(\{a, c, g, h, i, j, k, l, m\}) = 9$

(d) $n(C - A) = n(\{e, g, h\}) = 3$

CLASSIC CONUNDRUM   A Problem from the Dark Ages (page 153)

Give each of two of the sons five full flasks and five empty flasks and the other son ten half empty flasks. This gives each son ten half-flasks of contents, and ten flasks.

Chapter 2 Review Exercises (page 155)

1. (a) $S = \{4, 9, 16, 25\}$
   $P = \{2, 3, 5, 7, 11, 13, 17, 19, 23\}$
   $T = \{2, 4, 8, 16\}$

   (b) $\overline{P} = \{4, 6, 8, 9, 10, 12, 14, 15, 16, 18, 20, 21, 22, 24, 25\}$
   $S \cap T = \{4, 16\}$
   $S \cup T = \{2, 4, 8, 9, 16, 25\}$
   $S - T = \{9, 25\}$

2.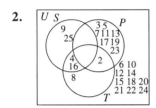

3. $S = \{\text{standard, legal}\}$, $C = \{\text{grey, tan, blue, yellow}\}$
   $S \times C = \{(\text{standard, grey}), (\text{standard, tan}), (\text{standard, blue}), (\text{standard, yellow}), (\text{legal, grey}),$
   $(\text{legal, tan}), (\text{legal, blue}), (\text{legal, yellow})\}$

4. $n(S) = 3$, $n(T) = 6$,
   $n(S \cup T) = n(\{s, e, t, h, o, r, y\}) = 7$,
   $n(S \cap T) = n(\{e, t\}) = 2$,
   $n(S - T) = n(\{s\}) = 1$,
   $n(T - S) = n(\{h, o, r, y\}) = 4$

5.
1	4	9	16	25	36	49	64	81	100
↕	↕	↕	↕	↕	↕	↕	↕	↕	↕
a	b	c	d	e	f	g	h	i	j

6. There is a one-to-one correspondence between the set of cubes and a proper subset. For example,

1	8	27	64	125	...
↕	↕	↕	↕	↕	
1	64	729	4096	15,625	...

7. (a) Suppose $A = \{a, b, c, d, e\}$ and $B = \{\blacksquare, \star\}$. Then $n(A) = 5$, $n(B) = 2$, $A \cap B = \emptyset$ and $n(A \cup B) = n(\{a, b, c, d, e, \blacksquare, \star\}) = 7$.

   (b)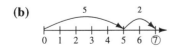

8. (a) Commutative property of addition:
   $7 + 3 = 3 + 7$

   (b) Associative property of addition:
   $3 + (2 + 6) = (3 + 2) + 6$

   (c) Additive identity property of zero:
   $7 + 0 = 7$

9. (a)

   (b)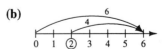

10. Answers may vary. Since $36 = 3 \times 3 \times 4$, one possibility is $6" \times 6" \times 8"$.

11. Since $92 \div 12 = 7$ R 8, there are eight rows (7 full rows, and a partial row of 8 soldiers in the back).

12. (a)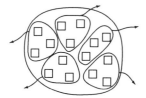

(b)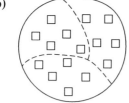

(c)

13. (a) Transitive only

(b) Reflexive, symmetric, and transitive (assuming that each person lives in exactly one country).

(c) Symmetric only. (This relation is not reflexive because some people cannot speak any language. It is not transitive because some people are bilingual—it is possible that Alan and Betty speak English, Betty and Charles speak Korean, but Alan and Charles have no common language.)

14. The relation is reflexive since any element belongs to the same subset as itself. The relation is symmetric since, if x is in the same subset as y, then y is in the same subset as x. The relation is transitive since, if x is in the same subset as y and y is in the same subset as z (and the subsets are mutually exclusive), then x must be in the same subset as z. Thus, $\square$ is an equivalence relation on S with equivalence classes A, B, and C.

15. (a) $f(3) = 2 \cdot 3 \cdot (3 - 3) = 0$
$f(0.5) = 2 \cdot 0.5 \cdot (0.5 - 3) = -2.5$
$f(-2) = 2 \cdot (-2) \cdot (-2 - 3) = 20$

(b) $x = 0$ or $x = 3$

16. (a)

(b) (Number of paths with n reflections)
= (number of paths with $n - 1$ reflections) + (number of paths with $n - 2$ reflections).
That is, the number of paths with n reflections is the Fibonacci number F_{n+2}.
To see why this is so, let P_n be the number of paths with n reflections. There are P_{n-1} such paths that begin by reflecting off the bottom surface. If the path begins by reflecting off the middle surface, it must have its second bounce off the top surface—so there are P_{n-2} of these paths. Therefore, $P_n = P_{n-1} + P_{n-2}$.

Chapter 2 Test (page 156)

1. 15th—ordinal
   1040—nominal
   $253—cardinal

2. (a)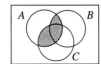

   (b)

   (c)

3. Since $A \cap B = A$, we know that $A \subseteq B$. Therefore $A - B = \emptyset$.

4. (a) $n(A \cup B)$
   $= n(\{w, h, o, l, e, n, u, m, b, r\}) = 10$

   (b) $n(B - C) = n(\{n, u, m, b\}) = 4$

   (c) $n(A \cap C) = n(\{O, e\}) = 2$

   (d) $n(A \times C) = n(A) \times n(C) = 5 \times 4 = 20$

5. (a) Associative property of addition

   (b) Distributive property of multiplication over addition

   (c) Additive identity property of zero

   (d) Associative property of multiplication

6. (a) Since 5 gallons = 20 quarts
   = 640 ounces, the number of bottles is
   $640 \div 10 = 64$.

   (b) Grouping

7. Since $n(A \cup B) = 21 = 12 + 14 - 5$, there must be 5 elements that are in both sets. Thus, $n(A \cap B) = 5$, and $n(\overline{A \cap B}) = 26 - 5 = 21$.

8. The possible values are the natural number factors of 21: 1, 3, 7, 21.

9. (a) $4 \times 2 = 8$

   (b) $12 \div 3 = 4$

   (c) $5 \cdot (9 + 2) = 5 \cdot 9 + 5 \cdot 2$

   (d) $10 - 4 = 6$

10. (a) Yes, because $2^a \cdot 2^b = 2^{a+b}$.

    (b) No, for example, $2 \in S$ and $4 \in S$ but $4 + 2 \notin S$.

11. (a) [diagram: 2×5 grid with divider = 2×3 grid + 2×2 grid]

    (b) [diagram: 2×5 grid = 5×2 grid]

12. (a) Number line

    (b) Comparison

    (c) Missing addend

13. (a) Let P, Q, and R be points in the plane. The relation is reflexive since $OP = OP$. The relation is symmetric since, if $OP = OQ$, then $OQ = OP$. Finally the relation is transitive since, if $OP = OQ$ and $OQ = OR$, then $OP = OR$. Thus, □ is an equivalence relation on S.

    (b) Each equivalence class is the set of points on a circle with center at the origin O.

14. The relation is reflexive and transitive. It is not symmetric because, for example, Los Angeles is no farther from the equator than Seattle, but it is *not* true that Seattle is no farther from the equator than Los Angeles.

15. (a)

    (b)

    (c)

    (d)

16. Find the function values:
    $f(1) = 1^2 - 4 \cdot 1 + 5 = 2$
    $f(2) = 2^2 - 4 \cdot 2 + 5 = 1$
    $f(3) = 3^2 - 4 \cdot 3 + 5 = 2$
    $f(4) = 4^2 - 4 \cdot 4 + 5 = 5$
    $f(5) = 5^2 - 4 \cdot 5 + 5 = 10$
    The range is $\{1, 2, 5, 10\}$.

17. (a) There are 5 paths: 1–3–5, 1–2–3–5, 1–3–4–5, 1–2–4–5, and 1–2–3–4–5.

    (b) The bee can enter cell $n + 2$ from either cell $n + 1$ or cell n. If there are $F(n + 1)$ ways to get to cell $n + 1$ and $F(n)$ ways to get to cell n, then the number of ways to get to cell $n + 2$ is $F(n + 1) + F(n)$.

    (c) $F(n)$ is the *n*th Fibonacci number, so $F(12) = 144$.

Chapter 3

JUST FOR FUN **For Careful Readers (page 167)**

1. None; it's a hole!

2. Three inches, the difference between the radii of the two circles. The needle does not rotate.

3. A quarter and a penny. One was not a quarter but the other was.

Problem Set 3.1 (page 173)

1. (a) $2 \cdot 1000 + 1 \cdot 100 + 3 \cdot 10 + 7 = 2137$

   (c) $100{,}000 + 2 \cdot 10{,}000 + 3 \cdot 100 + 1 \cdot 10 = 120{,}310$

   (e) $500 + 100 + 90 + 5 + 2 = 697$

   (g) $2 \cdot 7200 + 6 \cdot 360 + 18 \cdot 20 + 0 \cdot 1 = 16{,}920$

2. (a) $11 = 1 \cdot 10 + 1 \cdot 1 =$

3. (a) $9 = 10 - 1 = \text{IX}$

4. (a) $12 = 12 \cdot 1$

11. $452 = 4 \cdot 100 + 5 \cdot 10 + 2 \cdot 1$

    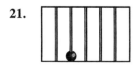

15. First trade 10 of your units for a strip to get 3 mats, 25 strips, and 3 units. Then trade 20 strips for 2 mats to get 5 mats, 5 strips, and 3 units.

18. (a) $2 \cdot 5 + 4 = 14$

    (c) $2 \cdot 5 + 3 = 13$

    (e) $1 \cdot 5^3 + 3 \cdot 5^2 + 3 \cdot 5 + 2$
    $= 1 \cdot 125 + 3 \cdot 25 + 3 \cdot 5 + 2 = 217$

20. (a) 413

21.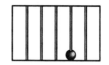

22. (a) $5 = 1 \cdot 5 + 0$

    (c) $10 = 2 \cdot 5 + 0$

    (e) $32 = 1 \cdot 5^2 + 1 \cdot 5 + 2$

    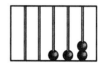

    (g) $125 = 1 \cdot 5^3 + 0 \cdot 5^2 + 0 \cdot 5 + 0$

    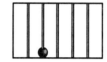

23. (a) The "fives" wire

25. (a) Yes. Any whole number can be represented. No matter what number is represented, one can always be added by bringing over one more bead on the rightmost wire and then working from right to left following the rule: If there are 5 beads showing on any wire, then they shall be moved to the back and one bead brought forward on the wire immediately on the left.

27. (a) $A \cup B = \{1, 2, 3, 4, 5, 7, 9\}$

    (c) $\overline{A} = \{2, 4, 6, 8\}$

    (e) $A \cap (B \cup C)$
    $= \{1, 3, 5, 7, 9\} \cap \{1, 2, 3, 4, 6, 8\}$
    $= \{1, 3\}$

    (g) $n(A) = 5, n(C) = 4, n(A \cup C) = 9$

    (i) In (g), $n(A \cup C) = n(A) + n(C)$ because A and C are disjoint—that is, $A \cap C = \varnothing$. (The formula $n(A \cup C) = n(A) + n(C) - n(A \cap C)$ is correct even here, because $n(A \cap C) = 0$.) In (h), the sets A and B have two elements in common (1 and 3). The sum $n(A) + n(B)$ counts each of these elements twice, so it is necessary to subtract $n(A \cap B)$ to get $n(A \cup B)$.

28. (a) Distributive property of multiplication over addition

    (c) Commutative property of addition

29. (a) $10 - 3 = 7$, $10 - 7 = 3$

Problem Set 3.2 (page 179)

1.

Base Ten	Base 5
0	0
1	1
2	2
3	3
4	4
$5 = 1 \cdot 5 + 0$	10
$6 = 1 \cdot 5 + 1$	11
$7 = 1 \cdot 5 + 2$	12
$8 = 1 \cdot 5 + 3$	13
$9 = 1 \cdot 5 + 4$	14
$10 = 2 \cdot 5 + 0$	20
$11 = 2 \cdot 5 + 1$	21
$12 = 2 \cdot 5 + 2$	22
$13 = 2 \cdot 5 + 3$	23
$14 = 2 \cdot 5 + 4$	24
$15 = 3 \cdot 5 + 0$	30
$16 = 3 \cdot 5 + 1$	31
$17 = 3 \cdot 5 + 2$	32
$18 = 3 \cdot 5 + 3$	33
$19 = 3 \cdot 5 + 4$	34
$20 = 4 \cdot 5 + 0$	40
$21 = 4 \cdot 5 + 1$	41
$22 = 4 \cdot 5 + 2$	42
$23 = 4 \cdot 5 + 3$	43
$24 = 4 \cdot 5 + 4$	44
$25 = 1 \cdot 5^2 + 0 \cdot 5 + 0$	100

5. (a) $413_{\text{five}} = 4 \cdot 5^2 + 1 \cdot 5 + 3 = 108$

   (c) $10_{\text{five}} = 1 \cdot 5 + 0 = 5$

   (e) $1000_{\text{five}} = 1 \cdot 5^3 + 0 \cdot 5^2 + 0 \cdot 5 + 0 = 125$

6. (a) $413_{\text{six}} = 4 \cdot 6^2 + 1 \cdot 6 + 3 = 153$

   (c) $10_{\text{six}} = 1 \cdot 6 + 0 = 6$

   (e) $1000_{\text{six}} = 1 \cdot 6^3 + 0 \cdot 6^2 + 0 \cdot 6 + 0 = 216$

7. (a) $413_{\text{twevle}} = 4 \cdot 12^2 + 1 \cdot 12 + 3 = 591$

   (c) $10_{\text{twelve}} = 1 \cdot 12 + 0 = 12$

   (e) $1000_{\text{twelve}} = 1 \cdot 12^3 + 0 \cdot 12^2 + 0 \cdot 12 + 0 = 1728$

8. (a) $362 = 2 \cdot 125 + 112$
   $= 2 \cdot 125 + 4 \cdot 25 + 12$
   $= 2 \cdot 125 + 4 \cdot 25 + 2 \cdot 5 + 2$
   $= 2422_{\text{five}}$

   (c) $5 = 1 \cdot 5 + 0 = 10_{\text{five}}$

9. (a) $342 = 1 \cdot 216 + 126$
   $= 1 \cdot 216 + 3 \cdot 36 + 18$
   $= 1 \cdot 216 + 3 \cdot 36 + 3 \cdot 6 + 0$
   $= 1330_{\text{six}}$

   (c) $6 = 1 \cdot 6 + 0 = 10_{\text{six}}$

10. (a) $342_{\text{six}} = 3 \cdot 6^2 + 4 \cdot 6 + 2$
    $= 134$
    $= 1 \cdot 125 + 9$
    $= 1 \cdot 125 + 0 \cdot 25 + 1 \cdot 5 + 4$
    $= 1014_{\text{five}}$

    (c) $41_{\text{six}} = 4 \cdot 6 + 1$
    $= 25$
    $= 1 \cdot 25 + 0 \cdot 5 + 0$
    $= 100_{\text{five}}$

11. (a) $2743 = 1 \cdot 1728 + 1015$
    $= 1 \cdot 1728 + 7 \cdot 144 + 7$
    $= 1 \cdot 1728 + 7 \cdot 144 + 0 \cdot 12 + 7$
    $= 1707_{\text{twelve}}$

    (c) $144 = 1 \cdot 144 + 0 \cdot 12 + 0 = 100_{\text{twelve}}$

12. (a)

one thousand twenty-fours	five hundred twelves	two hundred fifty-sixes	one hundred twenty-eights	sixty-fours	thirty-twos	sixteens	eights	fours	twos	units
1024	512	256	128	64	32	16	8	4	2	1
2^{10}	2^9	2^8	2^7	2^6	2^5	2^4	2^3	2^2	2^1	2^0

(c) (i) $24 = 16 + 8 = 1 \cdot 16 + 1 \cdot 8 + 0 \cdot 4 + 0 \cdot 2 + 0 \cdot 1 = 11000_{two}$

(ii) $18 = 16 + 2 = 1 \cdot 16 + 0 \cdot 8 + 0 \cdot 4 + 1 \cdot 2 + 0 \cdot 1 = 10010_{two}$

(iii) $2 = 1 \cdot 2 + 0 \cdot 1 = 10_{two}$

(iv) $8 = 1 \cdot 8 + 0 \cdot 4 + 0 \cdot 2 + 0 \cdot 1 = 1000_{two}$

14. (a) Note that this is similar to adding 99,999,999 + 1 in base ten.

$$11,111,111_{two}$$
$$+ \quad 1_{two}$$
$$\overline{100,000,000_{two}}$$

16. (a) Each 3-digit sequence of 0s and 1s must begin with either 0 or 1. The set of all those beginning with 0 can be obtained by appending a 0 to the left end of each of the different 2-digit sequences. In a similar way, we obtain all 3-digit sequences beginning with 1. Thus, appending a 0 and then a 1 to the left end of each 2-digit sequence, we obtain 000, 100, 010, 110, 001, 101, 011, and 111.

(c) Because of the way they were obtained, it is clear that there are twice as many 4-digit sequences as 3-digit sequences. Since the 5-digit sequences can be obtained from the 4-digit sequences in the same way, there are twice as many (i.e., 32) 5-digit sequences.

17. (a) They are 0, 1, 2, 3, 4, 5, 6, and 7 (but not in that order)—namely, the numbers that can be written with three or fewer digits in base 2.

(c) The whole numbers from 0 to $2^n - 1$. There are 2^n of these whole numbers, each with a different *n*-digit base 2 representation corresponding to one of the *n*-digit sequences of 0s and 1s.

20. (a) $51 \div 3 = 17$, $51 \div 17 = 3$

21. Answers will vary.

(a) $11 \times 31 = 341$, $341 \div 31 = 11$

23. (a) $n + 6$

JUST FOR FUN   **What's the Difference? (page 192)**

$$\begin{array}{r} 91 \\ -19 \\ \hline 72 \end{array} \quad \begin{array}{r} 95 \\ -59 \\ \hline 36 \end{array} \quad \begin{array}{r} 42 \\ -24 \\ \hline 18 \end{array} \quad \begin{array}{r} 62 \\ -26 \\ \hline 36 \end{array} \quad \begin{array}{r} 74 \\ -47 \\ \hline 27 \end{array}$$

$$\begin{array}{r} 61 \\ -16 \\ \hline 45 \end{array} \quad \begin{array}{r} 82 \\ -28 \\ \hline 54 \end{array} \quad \begin{array}{r} 81 \\ -18 \\ \hline 63 \end{array} \quad \begin{array}{r} 32 \\ -23 \\ \hline 9 \end{array} \quad \begin{array}{r} 54 \\ -45 \\ \hline 9 \end{array}$$

(a) All answers are evenly divisible by 9.

(b) Yes. The first digit in the answer is 1 less than the difference between the digits in the minuhend (the top number) in each subtraction problem and the second digit is 9 minus the first digit.

(c) In each case the sum of the digits is 9.

Problem Set 3.3 (page 195)

1. **(a)** $36 + 75 = 111$

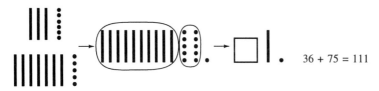

$36 + 75 = 111$

2. **(a)**
$$\begin{array}{r} 23 \\ +44 \\ \hline 7 \\ 60 \\ \hline 67 \end{array}$$

5. **(b)** First exchange one of the tens in 275 to 10 ones.

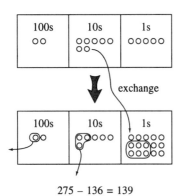

$275 - 136 = 139$

6. **(a)**
$$\begin{array}{r} 7\;5 \\ -3\;5 \\ \hline 4\;3 \end{array}$$

10. **(a)**

Base ten	Base 4
0	0
1	1
2	2
3	3
$4 = 1 \cdot 4 + 0$	10
$5 = 1 \cdot 4 + 1$	11
$6 = 1 \cdot 4 + 2$	12
$7 = 1 \cdot 4 + 3$	13
$8 = 2 \cdot 4 + 0$	20
$9 = 2 \cdot 4 + 1$	21
$10 = 2 \cdot 4 + 2$	22
$11 = 2 \cdot 4 + 3$	23
$12 = 3 \cdot 4 + 0$	30
$13 = 3 \cdot 4 + 1$	31
$14 = 3 \cdot 4 + 2$	32
$15 = 3 \cdot 4 + 3$	33

12. **(a)**
$$\begin{array}{r} \overset{1}{\;}231_{four} \\ +121_{four} \\ \hline 1012_{four} \end{array} \qquad \text{Check:} \begin{array}{r} 45 \\ +25 \\ \hline 70 \end{array}$$

(c)
$$\begin{array}{r} \overset{111}{1223}_{four} \\ +\;231_{four} \\ \hline 2120_{four} \end{array} \qquad \text{Check:} \begin{array}{r} 107 \\ +45 \\ \hline 152 \end{array}$$

(e) $\quad\overset{1}{3}2_{\text{four}}$ Check: $\quad 14$
$\quad\underline{+13_{\text{four}}}\qquad\qquad\underline{+\ 7}$
$\quad 111_{\text{four}}\qquad\qquad\ \ 21$

(g) $\quad\overset{1\ 10}{2}\overset{}{1}^{1}2_{\text{four}}$ Check: $\quad 38$
$\quad\underline{-\ 33_{\text{four}}}\qquad\qquad\underline{-15}$
$\quad\ 113_{\text{four}}\qquad\qquad\ \ 23$

13. (a) Work from the right column to the left column as shown:

    $\quad 6\ -\ -\ 3\qquad\overset{1}{6}\ -\ -\ 3\qquad\overset{1}{6}\ -\ 63$
    $\underline{+\ -\ 51\ -}\qquad\underline{+\ -\ 519}\qquad\underline{+\ -\ 519}$
    $\ \ -2282\qquad\ \ -2282\qquad\ \ -2282$

    $\quad\overset{1\ \ 1}{6}763\qquad\overset{1\ \ 1}{6}763$
    $\underline{+\ -\ 519}\qquad\underline{+\ \ 5519}$
    $\ \ -2282\qquad 12,282$

    (c) Work from right to left.
    $\quad\overset{1\ 1}{8}81$
    $\underline{+362}$
    $\ 1243$

    (e) Rewrite as an addition problem, then work from right to left.

    $\quad\ -15\ -\qquad\qquad\overset{1\ 1\ 1}{\ 2159}$
    $\underline{+\ 1843}\quad$ gives $\quad\underline{+\ 1843}$
    $\quad\ 4\ -\ -2\qquad\qquad\ \ 4002$

    Solution: $\quad 4002$
    $\qquad\qquad\underline{-1843}$
    $\qquad\qquad\ \ 2159$

14. (a) Work from right to left.
    $\quad\overset{1\ 1\ 1}{\ 2437}$
    $\ \ \ 281$
    $\underline{+\ 3476}$
    $\ \ 6194$

    (c) Work from right to left. Notice that the only way to get a 5 in the hundreds place of the sum is to "carry a 2," so the missing digits in the tens place of the addends must be 9s.
    $\quad 3891$
    $\quad 2493$
    $\underline{+5125}$
    $11,509$

15. (a) Rewrite as an addition problem, then work from right to left.

    $\quad\ 594\qquad\qquad\overset{1}{\ 594}$
    $\underline{+2\ -1}\quad$ gives $\quad\underline{+241}$
    $\quad\ -3\ -\qquad\qquad\ 835$

    Solution: $\quad 835$
    $\qquad\qquad\underline{-241}$
    $\qquad\qquad\ \ 594$

    (c) Rewrite as an addition problem, then work from right to left.

    $\quad\ 808\qquad\qquad\overset{1\ 1}{\ 808}$
    $\underline{+\ -5\text{-}4}\quad$ gives $\quad\underline{+6534}$
    $\quad\ 7\text{-}4\text{-}\qquad\qquad 7342$

    Solution: $\quad 7342$
    $\qquad\qquad\underline{-6534}$
    $\qquad\qquad\ \ 808$

16. (a) Five. Since 4s are used, the base is at least five. Since $1 + 4$ gives a 0 in the units column, the base is five.

    (c) Seven or greater. Since a 6 is used, the base is at least seven. No exchanges occur, so there is no other restriction.

    (e) Seven. Borrowing is required in the units column. If the base is b, the calculation is $13_b - 4_b = 6$, which means $b + 3 - 4 = 6$, or $b - 1 = 6$. Therefore, $b = 7$.

    (g) Twelve. Use the same reasoning as in (e): $b - 1 = 11$.

19. (a) $s = 8642 + 7531 + 1357 + 2468$
    $= 19{,}998$
    $1 \cdot s = 19{,}998\qquad 6 \cdot s = 119{,}988$
    $2 \cdot s = 39{,}996\qquad 7 \cdot s = 139{,}986$
    $3 \cdot s = 59{,}994\qquad 8 \cdot s = 159{,}984$
    $4 \cdot s = 79{,}992\qquad 9 \cdot s = 179{,}982$
    $5 \cdot s = 99{,}990\qquad 10 \cdot s = 199{,}980$

    (d) $14 \cdot s = 279{,}972$
    $23 \cdot s = 459{,}954$
    $32 \cdot s = 639{,}936$

    (f) Arithmetic progressions

21. (a) Note the exchange to the millions column.

    $\qquad\overset{1}{34{,}270{,}185}\ |\ 934{,}875$
    $\underline{+25{,}071{,}400}\ |\ 283{,}468$
    $\ \ 59{,}341{,}586\ |\ 218{,}343$

    Answer: $59{,}341{,}586{,}218{,}343$

22. (a)
$$\begin{array}{r|r} 65,421 & 534,784 \\ -24,131 & 243,677 \\ \hline 41,290 & 291,107 \end{array}$$

Answer: 41, 290,291,107

23. (a) $1001 \div 91 = 11$, $1001 \div 11 = 91$, $91 \times 11 = 1001$

24. Answers will vary.

(a) $143 \times 7 = 1001$, $1001 \div 143 = 7$

26. (a) $A \cup B \cup C = \{1, 2, 3, 4, 5, 6, 7, 8\}$
$A \cap B = \{2, 4\}$
$A \cap C = \{3, 4, 5\}$
$B \cap C = \{4, 6\}$
$A \cap B \cap C = \{4\}$

JUST FOR FUN   **What's the Sum? (page 203)**

$$\begin{array}{cccccc}
91 & 95 & 42 & 62 & 74 \\
+19 & +59 & +24 & +26 & +47 \\
\hline
110 & 154 & 66 & 88 & 121 \\
\\
61 & 82 & 81 & 32 & 54 \\
+16 & +28 & +18 & +23 & +45 \\
\hline
77 & 110 & 99 & 55 & 99
\end{array}$$

(a) All the divisions come out even.

(b) Yes. Note that each addition is a sum of two-digit numbers, where the second number is obtained by reversing the order of the digits of the first number. If the sum of the two digits is d, then the sum of the two numbers in question is $11 \cdot d$. Therefore, if $d < 10$, then the decimal representation of the sum in question is dd; that is, 66, or 55, or whatever. If d is the 2-digit number st, then the sum in question is $s(s + t)t$; that is, $1(1 + 4)4 = 154$, $1(1 + 1)1 = 121$, and so on.

Problem Set 3.4 (page 206)

1. (a) 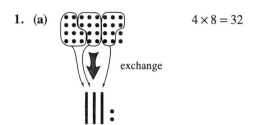 $4 \times 8 = 32$

3. (a) The number of hundreds in $30 \times 70 + 100$. You can see this by thinking of the calculation 30×274. The 100 comes from $30 \times 4 = 120 = 100 + 20$.

5. (a) Distributive property of multiplication over addition

(c) Associative property of addition

6.

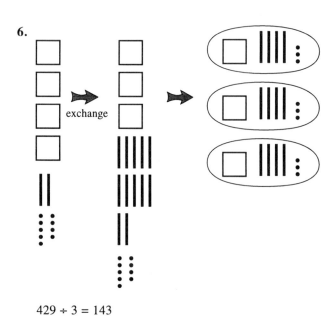

$429 \div 3 = 143$

9. (a)
$$4\overline{)27}\text{R }3$$
$$\underline{24}$$
$$3$$
$$6$$
   $27 = 4 \cdot 6 + 3$

10. (a)
$$351\overline{)7425}$$
   quotient digits: 21, 1, 20
   $\underline{7020}$
   405
   $\underline{351}$
   54

   The division checks because
   $7425 = 351 \cdot 21 + 54$.

11. (a) Follow these steps:
   $8 \div 5 = 1$ R 3
   $37 \div 5 = 7$ R 2
   $23 \div 5 = 4$ R 3

   $5\overline{)8^37^23}$ quotient 174 R 3

   Check: $873 = 5 \cdot 174 + 3$

14. (a) Reason as follows: $3 \times 3 = 9_{ten} = 14_{five}$, so write 4 and exchange 5 ones for 1 five. Then $3 \times 2 + 1 = 7_{ten} = 12_{five}$.

   $$\begin{array}{r} \overset{1}{}23_{five} \\ \times3_{five} \\ \hline 124_{five} \end{array}$$
   Check:
   $$\begin{array}{r} 13_{ten} \\ \times3_{ten} \\ \hline 39_{ten} \end{array}$$

15. (a) Base five:
   $4\overline{)231}$ quotient 31 R 2
   $\underline{22}$
   11
   $\underline{4}$
   2

   Check in base ten:
   $4\overline{)66}$ quotient 16 R 2
   $\underline{4}$
   26
   $\underline{24}$
   2

17. (a) They are the same because of the associative property of multiplication.
   $34 \cdot 54 = (17 \cdot 2) \cdot 54 = 17 \cdot (2 \cdot 54) = 17 \cdot 108$ since 2 evenly divides 34.

19. (a) Yes. This is simply a rearrangement of the rows in the usual algorithm. Thus, we usually write
   $$\begin{array}{r} 374 \\ \times23 \\ \hline 1122 \\ 748 \\ \hline 8602 \end{array}$$
   Here we multiply by 20 first and then 3 to obtain
   $$\begin{array}{r} 374 \\ \times23 \\ \hline 748 \\ 1122 \\ \hline 8602 \end{array}$$

(b)
$$\begin{array}{r} 285 \\ \times362 \\ \hline 855 \\ 1710 \\ 570 \\ \hline 103{,}170 \end{array}$$

20. (a)
   $7\overline{)1}$ 0 R 1
   $7\overline{)9}$ R 2
   $7\overline{)68}$ R 5
   $7\overline{)48^62}$ R 6
   $482_{ten} = 1256_{seven}$
   Check: $1 \cdot 7^3 + 2 \cdot 7^2 + 5 \cdot 7 + 6 = 482_{ten}$

(c)
   $2\overline{)1}$ 0 R 1
   $2\overline{)3}$ R 1
   $2\overline{)7}$ R 1
   $2\overline{)15}$ R 1
   $2\overline{)30}$ R 0
   $2\overline{)60}$ R 0
   $2\overline{)120}$ R 0
   $2\overline{)241}$ R 1
   $2\overline{)482}$ R 0
   $482_{ten} = 111100010_{two}$
   Check: $2^8 + 2^7 + 2^6 + 2^5 + 2^1 = 482_{ten}$

21. (a) The digits in the top number need to be written in decreasing order (to maximize the number of thousands, and so on). This eliminates all but five possibilities:

   $$\begin{array}{ccccc} 9753 & 9751 & 9731 & 9531 & \boxed{7531} \\ \times1 & \times3 & \times5 & \times7 & \times9 \\ \hline 9753 & 29{,}253 & 48{,}655 & 66{,}717 & 67{,}779 \end{array}$$

   The largest product is 7531×9.

24. (a) $375 \times 2432 = 912,000$

25. (a) $276,523 \div 511 \doteq 541.1409$
Quotient: 541
Remainder: $r \doteq 511 \cdot 0.1409 \doteq 72$
$276,523 \div 511 = 541 \text{ R } 72$
Check: $276,523 = 511 \cdot 541 + 72$

26. (a) $\quad 634 = 6 \text{ hundreds} + 3 \text{ tens} + 4 \text{ ones}$
$\underline{+\ 163 = 1 \text{ hundred } + 6 \text{ tens} + 3 \text{ ones}}$
$\qquad\quad 7 \text{ hundreds} + 9 \text{ tens} + 7 \text{ ones}$
$\qquad\quad = 797$

(c) $\quad 363 = 3 \text{ hundreds} + 6 \text{ tens} + 3 \text{ ones}$
$\underline{+\ 532 = 5 \text{ hundreds} + 3 \text{ tens} + 2 \text{ ones}}$
$\qquad\quad 8 \text{ hundreds} + 9 \text{ tens} + 5 \text{ ones}$
$\qquad\quad = 895$

(e) $\quad 725 = 7 \text{ hundreds} + 2 \text{ tens} + 5 \text{ ones}$
$\underline{-\ 413 = -(4 \text{ hundreds} + 1 \text{ ten} + 3 \text{ ones})}$
$\qquad\quad 3 \text{ hundreds} + 1 \text{ ten} + 2 \text{ ones}$
$\qquad\quad = 312$

27. (a) $\quad 374 = 3 \text{ hundreds} + 7 \text{ tens} + 4 \text{ ones}$
$\underline{+\ 483 = 4 \text{ hundreds} + 8 \text{ tens} + 3 \text{ ones}}$
$\qquad\quad 7 \text{ hundreds} + 15 \text{ tens} + 7 \text{ ones}$
$\qquad\quad = 8 \text{ hundreds} + 5 \text{ tens} + 7 \text{ ones}$
$\qquad\quad = 857$

(c) $\quad 724 = \qquad\qquad 7 \text{ hundreds} + 2 \text{ tens} + 4 \text{ ones}$
$\underline{+\ 532 = \qquad\qquad 5 \text{ hundreds} + 3 \text{ tens} + 2 \text{ ones}}$
$\qquad\qquad\qquad\quad 12 \text{ hundreds} + 5 \text{ tens} + 6 \text{ ones}$
$\qquad\quad = 1 \text{ thousand} + 2 \text{ hundreds} + 5 \text{ tens} + 6 \text{ ones}$
$\qquad\quad = 1256$

(e) $\quad 367 = 3 \text{ hundreds} + 6 \text{ tens} + 7 \text{ ones}$
$\underline{-249 = -(2 \text{ hundreds} + 4 \text{ tens} + 9 \text{ ones})}$
$\quad 367 = 3 \text{ hundreds} + 5 \text{ tens} + 17 \text{ ones}$
$\underline{-249 = -(2 \text{ hundreds} + 4 \text{ tens} + 9 \text{ ones})}$
$\qquad\quad 1 \text{ hundred} + 1 \text{ ten} + 8 \text{ ones}$
$\qquad\quad = 118$

28. (a) $\quad 213 = 2 \text{ twenty-fives} + 1 \text{ five } + 3 \text{ ones}$
$\underline{+\ 131 = 1 \text{ twenty-five } + 3 \text{ fives} + 1 \text{ one}}$
$\qquad\quad 3 \text{ twenty-fives} + 4 \text{ fives} + 4 \text{ ones}$
$\qquad\quad = 344_{\text{five}}$

(c) $\begin{aligned}142 &= 1 \text{ twenty-five } + 4 \text{ fives} + 2 \text{ ones}\\ +\;123 &= 1 \text{ twenty-five } + 2 \text{ fives} + 3 \text{ ones}\end{aligned}$

$\begin{aligned}&\;2 \text{ twenty-fives} + 6 \text{ fives} + 5 \text{ ones}\\ &= 2 \text{ twenty-fives} + 1 \text{ twenty-five } + 1 \text{ five } + 1 \text{ five} + 0 \text{ ones}\\ &= 3 \text{ twenty-fives} + 2 \text{ fives} + 0 \text{ ones}\\ &= 320_{\text{five}}\end{aligned}$

(e) $\begin{aligned}344 &= 3 \text{ twenty-fives} + 4 \text{ fives} + 4 \text{ ones}\\ -\;232 &= 2 \text{ twenty-fives} + 3 \text{ fives} + 2 \text{ ones}\end{aligned}$

$\begin{aligned}&= 1 \text{ twenty-five } + 1 \text{ five } + 2 \text{ ones}\\ &= 112_{\text{five}}\end{aligned}$

30. (a) $\begin{array}{r}\overset{1}{}34_{\text{five}}\\ +23_{\text{five}}\\ \hline 112_{\text{five}}\end{array}$

(c) $\begin{array}{r}\overset{2}{3}{}^{1}1\,2_{\text{five}}\\ -\;2\,1_{\text{five}}\\ \hline 2\,4\,1_{\text{five}}\end{array}$

(e) $\begin{array}{r}\overset{1}{}32_{\text{five}}\\ \times\;\;4_{\text{five}}\\ \hline 233_{\text{five}}\end{array}$

(g) $\begin{array}{r}12_{\text{five}}\;\;\text{R}13_{\text{five}}\\ 23_{\text{five}}\overline{)3\,4\,4_{\text{five}}}\\ 2\,3\\ \hline 1\,1\,4\\ 1\,0\,1\\ \hline 1\,3\end{array}$

Problem Set 3.5 (page 224)

1. Thought process may vary.

    (c) $\overparen{27 + 42} + 23$

    50, 90, 92

    (e) $48 \cdot 5 = (50 - 2) \cdot 5 = 250 - 10 = 240$

2. Thought process may vary.

    (c) $306 - 168 = 308 - 170, 130, 138$

    (e) $479 + 97 = 476 + 100, 576$

3. Thought process may vary.

    (a) $\begin{aligned}425 &+ 362\\ &= (400 + 20 + 5) + (300 + 60 + 2)\\ &= (400 + 300) + (20 + 60) + (5 + 2)\\ &= 700 + 80 + 7\end{aligned}$

    700, 780, 787

    (e) $\begin{aligned}3 &\cdot 342\\ &= 3 \cdot (300 + 40 + 2)\\ &= 3 \cdot 300 + 3 \cdot 40 + 3 \cdot 2\\ &= 900 + 120 + 6 = 1026\end{aligned}$

    900, 1000, 1026

4. (a) Round up because there is a 5 in the thousands place. 240,000

5. (a) 900

    (c) 27,000,000

6. (c) $600 - 500 < 678 - 431 < 700 - 400$
    $100 < 678 - 431 < 300$

    (e) $7000 \cdot 20 < 7403 \cdot 28 < 8000 \cdot 30$
    $140,000 < 7403 \cdot 28 < 240,000$

8. (a) $284 + 3046 \doteq 280 + 3050 = 300 + 3030 = 3330$

9. (a) $17,000 + 7000 + 12,000 + 2000 + 14,000 = 52,000$

    (c) $28,000 + 1000 + 2000 + 5000 + 12,000 = 48,000$

    (e) $21,000 - 8000 = 13,000$

10. (a) $3000 \cdot 30 = 90,000$

11. (a) $30,000 \div 40 = 750$

13. (a) $\dfrac{500 + 400}{300} = 3$

14. (a) Since $3 \cdot 7 = 21$, the last digit should be 1. 27,451 is correct.

16. Use guess and check.

   (a) $(24)(678) = 16,272$

   (c) $(2467)(8) = 19,736$

17. (a) The last digit is 6, so there must be 2 or 7 addends.
$(88) + 8 + 8 + 8 + 8 + 8 + 8 = 136$

   (c) The last digit is 8, so there must be 1 or 6 addends.
$(888) + 8 + 8 + 8 + 8 + 8 = 928$

18. Use guess and check. Some answers may vary.

   (a) $(844,422) \div 1 = 844,422$

   (c) $(84 \div (44 \div 22)) \div 1 = 42$

21. (a) Work from right to left.
```
  2 1 1
   2742
    415
   6943
  +2718
  12,818
```

22. (a) Rewrite as an addition problem, then work from right to left.

```
    91                  1
 + - 64 -      gives    91
   27 - 4             +2643
                       2743
```
Solution:
```
  2734
 −2643
    91
```

23. (a) The problem is:
```
    34 -
  ×  - -
    6 - 4
    - - 8
   - - 57 -
```
Since the first partial product is 600 plus, the units digit of the multiplier must be 2. Since the tens digit in the answer is 7 the missing digit in the first partial product must be 9. But then $694 \div 2 = 347$, so this is the multiplicand. Since the only digit that multiplies 7 to give 8 is 4, the multiplier must be 42. Thus, the solved problem is
```
    347
  ×  42
    694
   1388
  14,574
```

Problem Set 3.6 (page 238)

1. (a) $284 + 357 = 641$

   (c) $284 \cdot 357 = 101,388$

   (e) $781 - 35 + 24 = 770$

   (g) $781 - (35 - 24) = 770$

   (i) $861 - (423 + 201) = 237$

2. (a) $271 \cdot 365 = 98,915$

   (c) $1024 \div 16 \div 2 = 32$

3. (a) $(420 + 315) \div 15 = 49$

   (c) $(4441 + 2332) \div (220 + 301) = 13$

4. (a) [ON/AC] 29 [×], 841

   (c) [ON/AC] 2569 [−] 1480 [=] [√], 33

5. *Note*: Either calculation should give a result of 3.

   (a) [ON/AC] [(] 784 [√] [−] 91 [÷] 13 [)] [÷] [(] 8 [×] 49 [−] 11 [×] 35 [)] [=]

7. (a) $1831 - (17 \times 28) + 34$

13. Since $\frac{47-5}{3} = 14$, there are 15 terms in the sequence and the [=] key needs to be pressed 14 times.
[ON/AC] 5 [M+] [+] 3 [=] [M+] [=] [M+] [=] ⋯ [=] , [M+] [RM], 390

16. (a) This calculation evaluates $\dfrac{\left(\frac{\sqrt{5}+1}{2}\right)^3}{\sqrt{5}}$. It also stores $\dfrac{\sqrt{5}+1}{2}$ in memory for later use. After $\dfrac{\sqrt{5}+1}{2}$ is stored in memory, for example, $\dfrac{\left(\frac{\sqrt{5}+1}{2}\right)^5}{\sqrt{5}}$ results from RM y^x 5 = ÷ 5 √ =.

17. (a) $\dfrac{G-(1-G)}{\sqrt{5}} = 1$

   (c) $\dfrac{G^3-(1-G)^3}{\sqrt{5}} = 2$

   (e) $\dfrac{G^5-(1-G)^5}{\sqrt{5}} = 5$

18. Let $G = \dfrac{1+\sqrt{5}}{2}$, as in Problem 17.

   (a) $G + (1-G) = 1$

   (c) $G^3 + (1-G)^3 = 4$

   (e) $G^5 + (1-G)^5 = 11$

21. (b) Many calculators will run out of display room. You can calculate L_{39} by hand since $L_{39} = L_{37} + L_{38}$.
   $$\begin{array}{r} L_{37} = 54,018,521 \\ L_{38} = 87,403,803 \\ \hline L_{39} = 141,422,324 \end{array}$$

23. $17 + 18 = 35$; $35 - 17 = 18$; $35 - 18 = 17$

25. $11 \cdot 27 = 297$; $297 \div 27 = 11$; $297 \div 11 = 27$

27.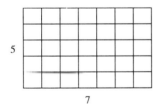

   $5 \cdot 7 = 35$

CLASSIC CONUNDRUM   **A Cryptarithm (page 241)**

$$\begin{array}{r} \text{FOOD} \\ +\ \text{FAD} \\ \hline \text{DIETS} \end{array}$$

Note that O is "oh" and not necessarily zero.

Since the sum of two different digits is at most 17 and the D in DIETS is a carry, it follows that D = 1 and S = 2. Also, since the carry to the last column comes from the fourth column, it must be the case that F = 9. Since I ≠ F, I must be 0. Then

$$\begin{array}{r} 9O O1 \\ +\ 9A1 \\ \hline 10ET2 \end{array}$$

If there is a carry from the second to the third column (where we count columns from the right), then O = E and this is a contradiction. Therefore, A + O = T.

The third column must carry to the fourth column, so O + 9 = 10 + E, which gives O = 1 + E. Substituting this value into A + O = T gives A + 1 + E = T.

The only digits from which to choose A, E, O, and T are 3, 4, 5, 6, 7, and 8. Since T ≤ 8, we know that A + 1 + E ≤ 8, or A + E ≤ 7. This means A and E must be 3 and 4 (in either order). But if E = 3, then O = 1 + E = 4, which contradicts the fact that O ≠ A.

Therefore, E = 4, A = 3, O = 1 + E = 5, and T = A + O = 8. There is only one solution:

$$\begin{array}{r} 9551 \\ +\ 931 \\ \hline 10482 \end{array}$$

Chapter 3 Review Exercises (page 243)

1. (a) 2000 + 300 + 50 + 3 = 2353

    (b) 8 · 7200 + 2 · 360 + 0 · 20 + 11 = 58,331

    (c) 1000 + 900 + 90 + 5 + 3 = 1998

2. 234,572 = 1 · 144,000 + 90,572
    = 1 · 144,000 + 12 · 7200 + 4172
    = 1 · 144,000 + 12 · 7200 + 11 · 360 + 212
    = 1 · 144,000 + 12 · 7200 + 11 · 360 + 10 · 20 + 12

3. Exchange 30 units for 3 strips, then exchange all 30 strips for 3 mats. The result is 8 mats, 0 strips, and 2 units.

4. (a) $1 \cdot 2^5 + 0 \cdot 2^4 + 1 \cdot 2^3 + 1 \cdot 2^2 + 0 \cdot 2 + 1 = 45_{\text{ten}}$

    (b) $3 \cdot 7^2 + 4 \cdot 7 + 6 = 181_{\text{ten}}$

    (c) $2 \cdot 12^2 + 10 \cdot 12 + 9 = 417_{\text{ten}}$

5. (a) 287 = 2 · 125 + 37
    = 2 · 125 + 1 · 25 + 12
    = 2 · 125 + 1 · 25 + 2 · 5 + 2
    = 2122_{five}

    (b) 287 = 1 · 256 + 31
    = 1 · 256 + 0 · 128 + 0 · 64 + 0 · 32 + 1 · 16 + 15
    = 1 · 256 + 0 · 128 + 0 · 64 + 0 · 32 + 1 · 16 + 1 · 8 + 7
    = 1 · 256 + 0 · 128 + 0 · 64 + 0 · 32 + 1 · 16 + 1 · 8 + 1 · 4 + 3
    = 1 · 256 + 0 · 128 + 0 · 64 + 0 · 32 + 1 · 16 + 1 · 8 + 1 · 4 + 1 · 2 + 1
    = 100011111_{two}

(c) $287 = 5 \cdot 49 + 42$
$= 5 \cdot 49 + 6 \cdot 7 + 0$
$= 560_{\text{seven}}$

6.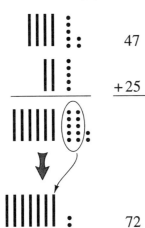

   47

   +25

   72

7. (a) $\begin{array}{r} 42 \\ +\ 54 \\ \hline 6 \\ 90 \\ \hline 96 \end{array}$

   (b) $\begin{array}{r} 47 \\ +\ 35 \\ \hline 12 \\ 70 \\ \hline 82 \end{array}$

   (c) $\begin{array}{r} 59 \\ +63 \\ \hline 12 \\ 110 \\ \hline 122 \end{array}$

8. (a)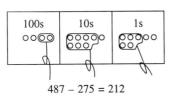

   $487 - 275 = 212$

   (b)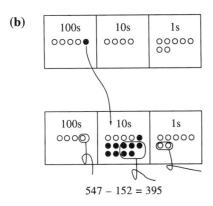

   $547 - 152 = 395$

9. (a) $\begin{array}{r} \overset{11}{2433}_{\text{five}} \\ +141_{\text{five}} \\ \hline 3124_{\text{five}} \end{array}$

   (b) $\begin{array}{r} 24\overset{3\ 1}{33}_{\text{five}} \\ -1\ 41_{\text{five}} \\ \hline 22\ 42_{\text{five}} \end{array}$

   (c)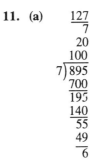

   $\begin{array}{r} \overset{3\,2}{}\\ \overset{1\,1}{243}_{\text{five}} \\ \times 42_{\text{five}} \\ \hline 1041 \\ 2132 \\ \hline 22411_{\text{five}} \end{array}$

10. (a) $\begin{array}{r} 357 \\ \times\ \ 4 \\ \hline 28 \\ 200 \\ 1200 \\ \hline 1428 \end{array}$

    (b) $\begin{array}{r} 642 \\ \times\ \ 27 \\ \hline 14 \\ 280 \\ 4200 \\ 40 \\ 800 \\ 12,000 \\ \hline 17,334 \end{array}$

11. (a) $\begin{array}{r} 127 \\ 7 \\ 20 \\ 100 \\ \hline 7\overline{)895} \\ 700 \\ \hline 195 \\ 140 \\ \hline 55 \\ 49 \\ \hline 6 \end{array}$

    (b) $\begin{array}{r} 79 \\ 9 \\ 70 \\ \hline 347\overline{)27,483} \\ 24,290 \\ \hline 3193 \\ 3123 \\ \hline 70 \end{array}$

12. (a) 5487 R 1
    $5\overline{)2\ 7^24^43^36}$

    (b) 4948 R 0
    $8\overline{)3\ 9^75^38^64}$

13. (a) $\overset{2}{\underset{\scriptscriptstyle 1}{}}$
    23_{five}
    $\times 42_{\text{five}}$
    $\overline{101}$
    202
    $\overline{2121_{\text{five}}}$

    (b) $\overset{2}{\underset{\scriptscriptstyle 1}{}}\ \overset{\ }{\underset{\scriptscriptstyle 1}{\ }}$
    2413_{five}
    $\times 332_{\text{five}}$
    $\overline{10331}$
    13244
    13244
    $\overline{2023221_{\text{five}}}$

14. | ~~42~~ | ~~35~~ |
    |---|---|
    | 21 | 70 |
    | ~~10~~ | ~~140~~ |
    | 5 | 280 |
    | ~~2~~ | ~~560~~ |
    | 1 | 1120 |
    | | 1470 |

15. (a) Round up because there is a 7 in the ten-thousands place.
    300,000

    (b) Round down because there is a 4 in the thousands place.
    270,000

    (c) Round up because there is a 5 in the hundreds place
    275,000

16. (a) $600 + 400 < 657 + 439 < 700 + 500$
    $1000 < 657 + 439 < 1200$

    (b) $600 - 500 < 657 - 439 < 700 - 400$
    $100 < 657 - 439 < 300$

    (c) $600 \cdot 400 < 657 \cdot 439 < 700 \cdot 500$
    $240,000 < 657 \cdot 439 < 350,000$

    (d) $1500 \div 30 < 1657 \div 23 < 1800 \div 20$
    $50 < 1657 \div 23 < 90$

17. (a) $700 + 400 = 1100$

    (b) $700 - 400 = 300$

    (c) $700 \cdot 400 = 280,000$

    (d) $2000 \div 20 = 100$

18. [ON/AC] 1444 [√] [−] 152 [÷] 19 [=] [+] [(] 2874 [−] 2859 [)] [=], 2

19. (a) Since $\dfrac{333-4}{7} = 47$, there are 48 terms and the [=] key must be pressed 47 times.
    [ON/AC] 4 [M+] [+] 7 [=] [M+] [=] [M+] ⋯ [=] [M+] [RM], 8088

    (b) Since $24,576 = 3 \cdot 2^{13}$, there are 14 terms and the [=] key must be pressed 13 times.
    [ON/AC] 3 [M+] [×] 2 [=] [M+] [=] [M+] ⋯ [=] [M+] [RM], 49,149

20. (a) Let $G = \dfrac{1+\sqrt{5}}{2}$.
    $7G \doteq 11$
    $11G \doteq 18$
    $18G \doteq 29$
    $29G \doteq 47$

    (b) The integer nearest $\left[\dfrac{1+\sqrt{5}}{2}\right] \cdot L_n = L_{n+1}$ for all integers n.

    (c) No. It does not hold for L_1, L_2 and L_3.

(d) The nearest integer to $\left[\frac{1+\sqrt{5}}{2}\right] \cdot L_n$ is L_{n+1} for $n \geq 4$. This seems to be true even for very large values of n, as the example of L_{34} and L_{35} illustrates.

Chapter 3 Test (page 244)

1. (a) $2 \cdot 3^4 + 1 \cdot 3^3 + 0 \cdot 3^2 + 2 \cdot 3 + 2 = 197_{\text{ten}}$

   (b) $3 \cdot 8^2 + 1 \cdot 8 + 7 = 207_{\text{ten}}$

   (c) $4 \cdot 5^3 + 2 \cdot 5^2 + 1 \cdot 5 + 3 = 558_{\text{ten}}$

2. (a) $281 = 2 \cdot 125 + 31$
$= 2 \cdot 125 + 1 \cdot 25 + 6$
$= 2 \cdot 125 + 1 \cdot 25 + 1 \cdot 5 + 1$
$= 2111_{\text{five}}$

   (b) $281 = 1 \cdot 256 + 25$
$= 1 \cdot 256 + 0 \cdot 128 + 0 \cdot 64 + 0 \cdot 32 + 1 \cdot 16 + 9$
$= 1 \cdot 256 + 0 \cdot 128 + 0 \cdot 64 + 0 \cdot 32 + 1 \cdot 16 + 1 \cdot 8 + 0 \cdot 4 + 0 \cdot 2 + 1$
$= 100011001_{\text{two}}$

   (c) $281 = 1 \cdot 144 + 137$
$= 1 \cdot 144 + 11 \cdot 12 + 5$
$= 1E5_{\text{twelve}}$

3. (a)
$$\begin{array}{r} \overset{1}{2}42_{\text{five}} \\ + 43_{\text{five}} \\ \hline 340_{\text{five}} \end{array}$$

   (b)
$$\begin{array}{r} 2\overset{1}{4}\overset{13}{2}_{\text{five}} \\ - 43_{\text{five}} \\ \hline 144_{\text{five}} \end{array}$$

   (c)
$$\begin{array}{r} \overset{3\,1}{}\\ \overset{2\,1}{}\\ 242_{\text{five}} \\ \times 43_{\text{five}} \\ \hline 1331 \\ 2123 \\ \hline 23111_{\text{five}} \end{array}$$

4. $74 + 48 = 122$

5. $39,485 = 5 \cdot 7200 + 3485$
    $= 5 \cdot 7200 + 9 \cdot 360 + 245$
    $= 5 \cdot 7200 + 9 \cdot 360 + 12 \cdot 20 + 5$

    ∷∷
    ∷
    ▬

6. Work from right to left.
    ```
      1 1
      2837
     +7224
     10,061
    ```

7. Rewrite as an addition problem and work from right to left.
    ```
      4 - 9 4              4694
     +3 5 - 2    gives    +3542
      - 2 3 -              8236
    ```
    Solution: 8236
              −3542
               4694

8. (a) Round down because a 3 is in the hundred-thousands place.
    3,000,000

    (b) Round up because a 7 is in the ten-thousands place.
    3,400,000

    (c) Round up because a 6 is in the thousands place.
    3,380,000.

    (d) Round up because a 5 is in the hundreds place.
    3,377,000

9. $400 + 60 + 300 + 40 + 4000 = 4800$

10. Again, the digits in each number need to be written in decreasing order. There are ten possibilities.

    ```
      531      731      751      753      931
       97       95       93       91       75
    51,507  69,445   69,843   68,523   69,825

      951      953      971      973      975
       73       71       53       51       31
    69,423  67,663   51,463   49,623   30,225
    ```

    The largest product is 751×93.

11. The digits in each number must increase from left to right (except 0, which must be in the second position from the left). This leaves ten possibilities:

    ```
      468      268      248      246      608
    ×  20    ×  40    ×  60    ×  80    ×  24
     9360   10,720   14,880   19,680   14,592

      408      406      208      206      204
    ×  26    ×  28    ×  46    ×  48    ×  68
    10,608   11,368    9568     9888   13,872
    ```

    The smallest product is 468×20.

12. (a) Let $G = 1 + \sqrt{2}$. Store this value in your calculator's memory.

    $f_1 = \dfrac{G}{\sqrt{8}} \doteq 0.854 \doteq 1$

    $f_2 = \dfrac{G^2}{\sqrt{8}} \doteq 2.061 \doteq 2$

    $f_3 = \dfrac{G^3}{\sqrt{8}} \doteq 4.975 \doteq 5$

    $f_4 = \dfrac{G^4}{\sqrt{8}} \doteq 12.010 \doteq 12$

(b) $f_5 \doteq 29$, $f_6 \doteq 70$

(c) The rule is $f_n = 2f_{n-1} + f_{n-2}$ for $n \geq 3$.

13. (a) 1 and 1
    9 and 9
    36 and 36
    100 and 100

    (b) 1, 3, 6, 10

    (c) Since the answers to (b) are the triangular numbers $t_n = \frac{n(n+1)}{2}$, conjecture that both $1^3 + 2^3 + \cdots + n^3$ and $(1+2+3+\cdots+n)^2 = \left(\frac{n(n+1)}{2}\right)^2$.

14. Since $\frac{123-3}{5} = 24$, there are 25 terms and the ⊟ key must be pressed 24 times.
    [ON/AC] 3 [M+] [+] 5 [=] [M+] [=] [M+] ⋯ [=] [M+] [RM], 1575

15. Since $1,171,875 = 3 \cdot 5^8$, there are 9 terms and the ⊟ key must be pressed 8 times.
    [ON/AC] 3 [M+] [×] 5 [=] [M+] [=] [M+] ⋯ [=] [M+] [RM], 1,464,843

16. Answers may vary. One possibility is similar to Algorithm 1 of Example 3.39:
    [ON/AC] 2 [M+] 5 [M+] [x⊂M] [M+] [x⊂M] [M+] ⋯

Chapter 4

JUST FOR FUN
A Problem of Punctuation
(page 260)

That that is, is; that that is not, is not.

Problem Set 4.1 (page 260)

1. (a)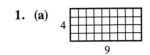

    $36 = 4 \cdot 9$
    4 divides 36.

3. (a) $0 \cdot 8 = 0$
    $1 \cdot 8 = 8$
    $2 \cdot 8 = 16$
    $3 \cdot 8 = 24$
    $4 \cdot 8 = 32$
    $5 \cdot 8 = 40$
    $6 \cdot 8 = 48$
    $7 \cdot 8 = 56$
    $8 \cdot 8 = 64$
    $9 \cdot 8 = 72$

5. Factor trees may vary, but the collection of prime numbers at the ends of the "branches" should agree in each case.

    (a)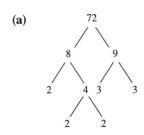

    (c)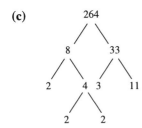

6. (a)
    ```
         5
    5)25
    2)50
    2)100
    7)700
    ```
    $700 = 7 \cdot 2 \cdot 2 \cdot 5 \cdot 5$

(c)
```
     2
3)6
3)18
5)90
5)450
```
$450 = 5 \cdot 5 \cdot 3 \cdot 3 \cdot 2$

7. (a) 1, 2, 3, 4, 6, 8, 12, 16, 24, 48

8. (a)
```
     3
2)6
2)12
2)24
2)48
```
$48 = 2 \cdot 2 \cdot 2 \cdot 2 \cdot 3 = 2^4 \cdot 3^1$

(c)
```
     5
5)25
5)125
3)375
3)1125
2)2250
```
$2250 = 2 \cdot 3 \cdot 3 \cdot 5 \cdot 5 \cdot 5 = 2^1 \cdot 3^2 \cdot 5^3$

9. (a) Yes. $28 = 2^2 \cdot 7^1$. Since 2^3 and 7^2 appear in the factorization of a, all the factors of 28 appear in a to at least as high a power.

   (c) The exponent on 2 is $3 - 2 = 1$, since 2^3 appears in a and 2^2 appears in b. Likewise, the exponent on 3 is $1 - 1 = 0$, and the exponent on 7 is 2 (unchanged from a). Thus $\frac{a}{b} = 2^1 \cdot 3^0 \cdot 7^2 = 2^1 \cdot 7^2 = 98$. (This may be written as $\frac{a}{b} = \frac{2^3 \cdot 3^1 \cdot 7^2}{2^2 \cdot 3^1}$
   $= 2^{3-2} \cdot 3^{1-1} \cdot 7^2 = 2^1 \cdot 7^2$.)

12. No. If $n = ab$, it need only be true that *one* of a or b is less than or equal to $\sqrt{n}$. For example, $10 = 2 \cdot 5$ with 2 and 5 both primes. Yet $5 > \sqrt{10}$.

13. (a) True. $n \cdot 0 = 0$ for every natural number n.

    (c) True. $1 \cdot n = n$ for every natural number n.

(e) False. The definition of $b|a$ specifies that $b \neq 0$. This is because $0 \div 0 = q$ if, and only if, $0 \cdot q = 0$ for a *unique* integer q. However, this is true for *every* integer q.

15. Yes. If $p|bc$ then p appears in the prime factorization of bc, which is the product of the prime factorization of b and the prime factorization of c. Therefore, p appears in the prime factorization of b or c (or both).

18. If $m|n$, then $n = mk$ for some natural number k. Also, if $n|m$ then $m = nq$ for some natural number q. Hence, $n = mk = nqk$ so $qk = 1$. Since q and k are both natural numbers, $q = 1$ and $k = 1$. Thus, $m = n$.
An alternate approach is to realize that since $n = mk$ for some natural number k, we know that $k \geq 1$ and hence $m \leq n$. Similarly, $n \leq m$ because $n|m$. Since $m \leq n$ and $n \leq m$, we can conclude that $m = n$.

21. No. For example, $2|(3+5)$ but $2 \nmid 3$ and $2 \nmid 5$.

23. (a) $N_2|N_4$ because $1111 = 11 \cdot 101$.
$N_2|N_6$ because $111{,}111 = 11 \cdot 10{,}101$.
$N_2|N_8$ because $11{,}111{,}111 = 11 \cdot 1{,}010{,}101$.

25. (a) No, $111 \nmid 11{,}111$.

(c) Yes, $111{,}111{,}111{,}111{,}111$
$= 111 \cdot 1{,}001{,}001{,}001{,}001$.

26. (a) $F_{11} = F_9 + F_{10} = 34 + 55 = 89$
$F_{12} = F_{10} + F_{11} = 55 + 89 = 144$
$F_{13} = F_{11} + F_{12} = 89 + 144 = 233$
$F_{14} = F_{12} + F_{13} = 144 + 233 = 377$
$F_{15} = F_{13} + F_{14} = 233 + 377 = 610$
$F_{16} = F_{14} + F_{15} = 377 + 610 = 987$
$F_{17} = F_{15} + F_{16} = 610 + 987 = 1597$
$F_{18} = F_{16} + F_{17} = 987 + 1597 = 2584$
$F_{19} = F_{17} + F_{18} = 1597 + 2584 = 4181$
$F_{20} = F_{18} + F_{19} = 2584 + 4181 = 6765$

(b) Since $F_3 = 2$, the Fibonacci numbers divisible by F_3 are the Fibonacci numbers that are even: $F_3 = 2$, $F_6 = 8$, $F_9 = 34$, $F_{12} = 144$, $F_{15} = 610$, $F_{18} = 2584$, and so on. It appears that $F_3|F_n$ if, and only if, $3|n$. Indeed, since $F_3 = 2$ and the sum of two odd numbers is even and the sum of an even and an odd number is odd, it follows that the Fibonacci sequence follows the pattern odd, odd, even, odd, odd, even, and so on without end. Thus, every third number in the sequence is divisible by 2 and conversely.

29. Use program FACTOR INTEGER.

(a) $548 = 2^2 \cdot 137^1$

(c) $274 = 2^1 \cdot 137^1$

(e) 548 does not divide any other number given because only 274 has a factor of 137, but its power of 2 is smaller. 936 divides 45,864 because 45,864 has powers of 2, 3, and 13 that are at least as great: $2^3 \cdot 3^2 \cdot 13^1 | 2^3 \cdot 3^2 \cdot 7^2 \cdot 13^1$. It does not divide 548 or 274, because these numbers do not have 13 in their prime factorizations.
274 divides 548 because 548 has powers of 2 and 137 that are at least as great: $2^1 \cdot 137^1 | 2^2 \cdot 137^1$. It does not divide 936 or 45,864 because these numbers do not have 137 in their prime factorization. 45,864 does not divide any other number given because the others do not have 7 in their prime factorization.
The only answers are $274|548$ (or $2^1 \cdot 137^1 | 2^2 \cdot 137^1$) and $936|45{,}864$ (or $2^3 \cdot 3^2 \cdot 13^1 | 2^3 \cdot 3^2 \cdot 7^2 \cdot 13^1$).

30. Use program FACTOR INTEGER.

(a) $894{,}348 = 2^2 \cdot 3^3 \cdot 7^2 \cdot 13^2$

(c) $1{,}265{,}625 = 3^4 \cdot 5^6$

(e) The exponents in the prime power representation of a square are even. This is because when you square a number, you double each exponent in its prime power representation.

31. (a) False. The sum of two odd natural numbers is an even natural number.

(c) True. $a + b = b + a$ if a and b are any natural numbers, so the commutative property holds for any subset of the natural numbers.

(e) True. $a + (b + c) = (a + b) + c$ if a, b, and c are any natural numbers.

(g) True. $a(b + c) = ab + ac$ if a, b, and c are any natural numbers.

(i) True. 1 is an element of S.

32. (a) $2 + 99 \cdot 3 = 299$

33. (a) Each term is 4 more than its predecessor: 19, 23, 27.

   (c) These are the triangular numbers: 15, 21, 28.

   (e) Each number is twice its predecessor: 48, 96, 192.

34. (a) Since $\frac{155 - 2}{3} = 51$, there are 52 terms.
$$S = 2 + 5 + 8 + \cdots + 51$$
$$S = 51 + 48 + 45 + \cdots + 2$$
$$2S = 53 + 53 + 53 + \cdots + 53$$
$$= 52 \cdot 53$$
$$= 2756$$
$$S = \frac{2756}{2} = 1378$$

Problem Set 4.2 (page 269)

1. (a) 1554 is divisible by 2 and 3: Divisible by 2 because units digits is even.
   Divisible by 3 because $1 + 5 + 5 + 4 = 15$ and $3|15$.
   Not divisible by 5 because units digit is not 0 or 5.

   (c) 805 is not divisible by 2 because units digit is odd.
   Not divisible by 3 because $8 + 0 + 5 = 13$ and $3 \nmid 13$.
   Divisible by 5 because units digit is 5.

2. (a) 1554. This is the only number that is divisible by 2 and 3.

   (c) None. None of the numbers are divisible by 3 and 5.

3. (a) 539 is divisible by 7 and 11.
   $7|539$ because $539 = 7 \cdot 77$.
   $11|539$ because $(5 + 9) - 3 = 11$ and $11|11$.
   $13 \nmid 539$ because $539 = 13 \cdot 41 + 6$.

   (c) 834,197 is divisible by 7 and 13. Use the combined test: $834 - 197 = 637$.
   $7|637$ because $637 = 7 \cdot 91$.
   $11 \nmid 637$ because $(6 + 7) - 3 = 10$ and $11 \nmid 10$.
   $13|637$ because $637 = 13 \cdot 49$.

4. (a) 539. This is the only number that is divisible by 7 and 11.

   (c) None. None of the numbers are divisible by 11 and 13.

6. (a) True. $2^3 \cdot 3^1 \cdot 5^2 | 2^4 \cdot 3^5 \cdot 5^3 \cdot 7^2 \cdot 11^2$ because each exponent in $2^3 \cdot 3^1 \cdot 5^2$ is no greater than the corresponding exponent in $2^4 \cdot 3^5 \cdot 5^3 \cdot 7^2 \cdot 11^2$.

   (c) True. $7^1 \cdot 11^2 | 2^4 \cdot 3^5 \cdot 5^3 \cdot 7^2 \cdot 11^2$ because each exponent in $7^1 \cdot 11^2$ is no greater than the corresponding exponent in $2^4 \cdot 3^5 \cdot 5^3 \cdot 7^2 \cdot 11^2$.

   (e) False.
   $ab = (2^3 \cdot 3^1 \cdot 5^2) \cdot (2^2 \cdot 3^2 \cdot 5^1 \cdot 7^1)$
   $= (2^3 \cdot 2^2) \cdot (3^1 \cdot 3^2) \cdot (5^2 \cdot 5^1) \cdot 7^1$
   $= 2^5 \cdot 3^3 \cdot 5^3 \cdot 7^1$, and
   $2^5 \cdot 3^3 \cdot 5^3 \cdot 7^1 \nmid 2^4 \cdot 3^5 \cdot 5^3 \cdot 7^2 \cdot 11^2$
   because the exponent on 2 in $2^5 \cdot 3^3 \cdot 5^3 \cdot 7^1$ is greater than the exponent on 2 in $2^4 \cdot 3^5 \cdot 5^3 \cdot 7^2 \cdot 11^2$.

8. (a) For any palindrome with an even number of digits, the digits in the odd positions are the same as the digits in the even positions, but with the order reversed. Thus, the difference of the sums of the digits in the even and odd positions is zero, which is divisible by 11.

9. When using the combined test for divisibility by 7, 11, and 13 the digits of the number are broken up into 3-digit groups. If a number has form abc, abc, then the difference in the sums of the 3-digit numbers in odd positions and even positions will be zero, which is divisible by each of 7, 11, and 13.
   An alternate explanation is to note that abc, $abc = 1001 \cdot abc = 7 \cdot 11 \cdot 13 \cdot abc$, where abc denotes the number $100a + 10b + c$.

12. No. He or she could have made other kinds of errors that by chance resulted in the record being out of balance by an amount that is a

multiple of 9. For example, the teller could have made a 5¢ error and later a 4¢ error, resulting in a total error of 9¢.

13. (e) No. For example, $27 | 692,334$ but $11 \nmid 692,334$.

14. (a) $126 \div 9 = 14$
    $10,206 \div 9 = 1134$
    $1,002,006 \div 9 = 111,334$
    $100,020,006 \div 9 = 11,113,334$
    $10,000,200,006 \div 9 = 1,111,133,334$
    $1,000,002,000,006 \div 9$
    $\qquad = 111,111,333,334$

16. Suppose d divides b and c then $b = dm$ and $c = dn$ for some natural numbers m and n. But $a = b + c = dm + dn = d(m + n)$. So d also divides a. Now suppose d divides a and b. (This is the same as if d divided a and c.) Then $a = dq$ and $b = dr$ for some natural numbers q and r. If $a = b + c$, then $c = a - b = dq - dr = d(q - r)$. So d also divides c.

18. (a)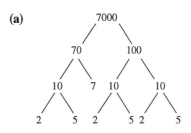

    $8064 = 2^7 \cdot 3^2 \cdot 7^1$

19. Factor trees may vary.

    (a)
    ```
                7000
               /    \
              70    100
             / \   / \  / \
            10  7 10   10
           / \   / \   / \
          2   5 2   5 2   5
    ```

    $7000 = 2^3 \cdot 5^3 \cdot 7^1$

21. $(2^3 \cdot 3^4 \cdot 5^1 \cdot 7^2) \div (2^1 \cdot 3^2 \cdot 7^1)$
    $= (2^3 \div 2^1) \cdot (3^4 \div 3^2) \cdot (5^1) \cdot (7^2 \div 7^1)$
    $= 2^2 \cdot 3^2 \cdot 5^1 \cdot 7^1$

JUST FOR FUN   Making a Chain (page 271)

Four. Open all four of the links in one section of chain. These can then be used to connect the remaining five sections of chain into a single chain of 24 links.

JUST FOR FUN   A Weighty Matter (page 274)

If the basketball weighs 21 ounces plus half its own weight, then 21 ounces must be half the weight of the ball. Therefore, the basketball weighs 42 ounces.

Problem Set 4.3 (page 284)

1. (a) Cut off a 9 by 9 square, then a 6 by 6 square, then a 3 by 3 square. A 3 by 3 square remains. The 3 by 3 square is the largest square that will tile the 9 by 15 rectangle.

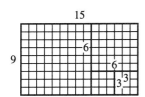

(b)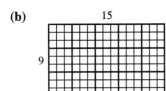

3. The rectangle at left in the next figure shows that GCD(4, 7) = 1, so we cannot simplify this problem as was done in Example 4.20. Therefore, use guess and check to determine that the 28 by 28 square is the smallest square obtainable. LCM(4, 7) = 28.

 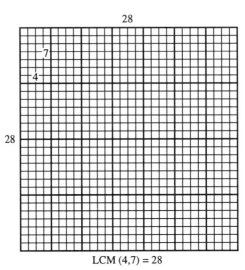

LCM (4,7) = 28

5. (a) Let D_{24} and D_{27} represent the sets of divisors of 24 and 27.
   $D_{24} = \{1, 2, 3, 4, 6, 8, 12, 24\}$
   $D_{27} = \{1, 3, 9, 27\}$
   $D_{24} \cap D_{27} = \{1, 3\}$
   GCD(24, 27) = 3

6. (a) Let M_{24} and M_{27} represent the set of multiples of 24 and 27.
   $M_{24} = \{0, 24, 48, 72, 96, 120, 144, 168, 192, 216, 240, \cdots\}$
   $M_{27} = \{0, 27, 54, 81, 108, 135, 162, 189, 216, 243, \cdots\}$
   $M_{24} \cap M_{27} = \{0, 216, \cdots\}$
   LCM(24, 27) = 216

7. (a) GCD(24, 27) · LCM(24, 27) = 3 · 216 = 648 = 24 · 27
   GCD(14, 22) · LCM(14, 22) = 2 · 154 = 308 = 14 · 22
   GCD(48, 72) · LCM(48, 72) = 24 · 144 = 3456 = 48 · 72

8. (a) For the GCD we choose the smaller of the two exponents with which each prime appears in r and s:
   $GCD(r, s) = 2^1 \cdot 3^1 \cdot 5^2 = 150$.
   For the LCM we choose the larger of the two exponents with which each prime appears in r and s:
   $LCM(r, s) = 2^2 \cdot 3^3 \cdot 5^3 = 13,500$.

10. (a) $a = 2^2 \cdot 3^1 \cdot 5^2 \cdot 7^0$
    $b = 2^1 \cdot 3^3 \cdot 5^1 \cdot 7^0$
    $c = 2^0 \cdot 3^2 \cdot 5^3 \cdot 7^1$
    For the GCD we choose the smallest of the three exponents with which each prime appears in a, b, and

c: $GCD(a, b, c) = 2^0 \cdot 3^1 \cdot 5^1 \cdot 7^0 = 15$. For the LCM we choose the largest of the three exponents with which each prime appears in a, b, and c: $LCM(a, b, c) = 2^2 \cdot 3^3 \cdot 5^3 \cdot 7^1 = 94,500$.

11. (a) $D_{18} = \{1, 2, 3, 6, 9, 18\}$
    $D_{24} = \{1, 2, 3, 4, 6, 8, 12, 24\}$
    $D_{12} = \{1, 2, 3, 4, 6, 12\}$
    $D_{18} \cap D_{24} \cap D_{12} = \{1, 2, 3, 6\}$
    $GCD(18, 24, 12) = 6$
    $M_{18} = \{0, 18, 36, 54, 72, 90, \cdots\}$
    $M_{24} = \{0, 24, 48, 72, 96, \cdots\}$
    $M_{12} = \{0, 12, 24, 36, 48, 60, 72, 84, \cdots\}$
    $M_{18} \cap M_{24} \cap M_{12} = \{0, 72, \cdots\}$
    $LCM(18, 24, 12) = 72$

13. (a) $GCD(24, 18) = 6$;
    $GCD(GCD(24, 18), 12) = GCD(6, 12)$
    $= 6$
    $LCM(24, 18) = 72$;
    $LCM(LCM(24, 18), 12) = LCM(72, 12)$
    $= 72$
    The final results are the same.

14. Note that the x rod measures the y rod if, and only if, $x|y$.

    (a) The 1, 3, and 9 rods. (The divisors of 9 are 1, 3, and 9.)

    (d) Any rods or trains with length 1, 2, 3, 6, 9, or 18. (The divisors of 18 are 1, 2, 3, 6, 9, and 18.)

15. (a) The 4-rods can measure trains of length 4, 8, 12, 16, $\cdots$. The 6-rods can measure trains of length 6, 12, 18, $\cdots$. The shortest train that can be measured by 4-rods and by 6-rods has length 12.

    (b) $LCM(4, 6) = 12$

19. (a) $F_{12} \div F_6 = 144 \div 8 = 18$
    $F_{18} \div F_9 = 2584 \div 34 = 76$
    $F_{30} \div F_{15} = 832,040 \div 610 = 1364$

    (c) No. $F_{19} = 4181 = 37 \cdot 113$. The correct statement is: If n is prime, then F_n does not have any factors that are Fibonacci numbers, other than itself and 1.

    (h) The conjecture predicts that $GCD(F_{16}, F_{20}) = F_{GCD(16, 20)}$ $= F_4 = 3$, so the conjecture would be false if $GCD(F_{16}, F_{20}) = 4$.

21. (a) 1224 seconds. Since $72 = 2^3 \cdot 3^2$ and $68 = 2^2 \cdot 17^1$,
    $LCM(72, 68) = 2^3 \cdot 3^2 \cdot 17^1 = 1224$, so the first time that Hi and Sarah return to the starting point simultaneously is after 1224 seconds. To confirm that Hi has lapped Sarah exactly once at this time, it is necessary to note that $1224 \div 72 = 17$ and $1224 \div 68 = 18$, so Sarah has completed 17 laps and Hi has completed 18.

23. (a) $220 = 2^2 \cdot 5^1 \cdot 11^1$
    $264 = 2^3 \cdot 3^1 \cdot 11^1$
    $275 = 5^2 \cdot 11^1$
    $LCM(220, 264) = 2^3 \cdot 3^1 \cdot 5^1 \cdot 11^1 = 1320$
    $LCM(220, 275) = 2^2 \cdot 5^2 \cdot 11^1 = 1100$
    $LCM(264, 275) = 2^3 \cdot 3^1 \cdot 5^2 \cdot 11^2 = 6600$

24. Use TILERECTANGLE.

    (a) $GCD(189, 294) = 21$

25. Use TILESQUARE.

    (a) $LCM(189, 294) = 2646$

26. Use DIVISORSETS.

    (a) $GCD(45, 48) = 3$

27. Use MULTIPLESETS.

    (a) $LCM(45, 48) = 720$

28. (a) $205,800 = 2^3 \cdot 3^1 \cdot 5^2 \cdot 7^3$
$31,460 = 2^2 \cdot 5^1 \cdot 11^2 \cdot 13^1$
$25,840 = 2^4 \cdot 5^1 \cdot 17^1 \cdot 19^1$

30. (a) $A \cap (B \cup C) = \{a, b, c, d, e, f, g\} \cap \{a, b, c, d, e, g\}$
$= \{a, b, c, d, e, g\}$
$(A \cap B) \cup (A \cap C) = \{a, c, d, e, g\} \cup \{a, b, c, d\}$
$= \{a, b, c, d, e, g\}$

(c) $\overline{A \cup B} = \overline{\{a, b, c, d, e, f, g\}} = \{h, i, j, k, l, m, n, o, p, q, r, s, t, u, v, w, x, y, z\}$
$\overline{A} \cap \overline{B} = \{h, i, j, k, l, m, n, o, p, q, r, s, t, u, v, w, x, y, z\}$
$\cap \{b, f, h, i, j, k, l, m, n, o, p, q, r, s, t, u, v, w, x, y, z\}$
$= \{h, i, j, k, l, m, n, o, p, q, r, s, t, u, v, w, x, y, z\}$

31. (a) Since the largest value of $n^2 - 81n + 1681$ is 1601, and $\sqrt{1601} \doteq 40.01$, we can check that these values are prime by checking for divisibility by 2, 3, 5, 7, 11, 13, 17, 19, 23, 29, 31, and 37.

n	$n^2 - 81n + 1681$	prime?
1	1601	yes
2	1523	yes
3	1447	yes
4	1373	yes
5	1301	yes

(c) No. If a conjecture is true for several cases, it does not mean it is true for *every* case.

(e) The conjecture may seem more probable, but we still cannot say if it is *always* true.

32. (a) No. If $p|a$ and $p \nmid b$ then $p \nmid (a+b)$, by Exercise 20 of Problem Set 4.1. An alternate explanation is to observe that, by the division algorithm, the remainder after dividing by 2, 5, or 7 will be 1.

33. (a) No. If $p|a$ and $p \nmid b$ then $p \nmid (a+b)$.

Problem Set 4.4 (page 304)

1. (a) $5 + 9 = 14 = 1 \cdot 12 + 2$.
Therefore, $5 +_{12} 9 = 2$.

(c) $8 + 4 = 12 = 1 \cdot 12 + 0$.
Therefore, $8 +_{12} 4 = 12$.

(e) $10 + 10 = 20 = 1 \cdot 12 + 8$.
Therefore, $10 +_{12} 10 = 8$.

2. (a) $3 + 4 = 7 = 1 \cdot 5 + 2$.
Therefore, $3 +_5 4 = 2$.

(c) $2 + 7 = 9 = 0 \cdot 10 + 9$.
Therefore, $2 +_{10} 7 = 9$.

(e) $12 + 10 = 22 = 1 \cdot 16 + 6$.
Therefore, $12 +_{16} 10 = 6$.

3. (a)

$9 -_{12} 7 = 2$

(c)

$5 -_{12} 9 = 8$

(e)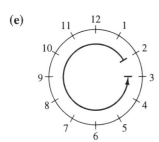

$2 -_{12} 11 = 3$

4. Find each additive inverse by subtracting each number from 12. (Exception: the additive inverse of 12 is 12.)

   (a) $12 - 7 = 5$

   (c) $12 - 9 = 3$

5. (a) $12 - 7 = 5$, so add 5:
   $9 -_{12} 7 = 9 +_{12} 5 = 2$.

   (c) $12 - 9 = 3$, so add 3:
   $5 -_{12} 9 = 5 +_{12} 3 = 8$.

   (e) The additive inverse of 11 is 1, so add 1: $2 -_{12} 11 = 2 +_{12} 1 = 3$.

6. (a) $5 \times 7 = 35 = 2 \cdot 12 + 11$.
   Therefore, $5 \times_{12} 7 = 11$.

   (c) $12 \times 5 = 60 = 5 \cdot 12 + 0$.
   Therefore, $12 \times_{12} 5 = 12$.

   (e) $4 \times 6 = 24 = 2 \cdot 12 + 0$.
   Therefore, $4 \times_{12} 6 = 12$.

7. (a) Look for a 5 in the 7th row of Table 4.1. Since $7 \times_{12} 11 = 5$ and 11 is the *only* element a in T for which $7 \times_{12} a = 5$, $5 \div_{12} 7 = 11$.

   (c) Look for an 8 in the 4th row of Table 4.1. Since there are several 8's in this row ($4 \times_{12} 2 = 8$, $4 \times_{12} 5 = 8$, $4 \times_{12} 8 = 8$, and $4 \times_{12} 11 = 8$), $8 \div_{12} 4$ is undefined.

   (e) Look for a 9 in the 5th row of Table 4.1. Since $5 \times_{12} 9 = 9$ and 9 is the *only* element a of T for which $5 \times_{12} a = 9$, $9 \div_{12} 5 = 9$.

8. (a) GCD(1, 12) = 1
   GCD(2, 12) = 2
   GCD(3, 12) = 3
   GCD(4, 12) = 4
   GCD(5, 12) = 1
   GCD(6, 12) = 6
   GCD(7, 12) = 1
   GCD(8, 12) = 4
   GCD(9, 12) = 3
   GCD(10, 12) = 2
   GCD(11, 12) = 1
   GCD(12, 12) = 12
   1, 5, 7, and 11 are relatively prime to 12.

10. (a) $3 + 4 = 7 = 1 \cdot 5 + 2$.
    Therefore, $3 +_5 4 = 2$.

    (c) $4 + 4 = 8 = 1 \cdot 5 + 3$.
    Therefore, $4 +_5 4 = 3$.

    (e) $2 \times 5 = 10 = 2 \cdot 5 + 0$.
    Therefore, $2 \times_5 5 = 5$.

    (g) $3 -_5 4 = 3 +_5 1 = 4$

    (i) $4 -_5 4 = 4 +_5 1 = 5$

    (k) $2 \div_5 5$ is undefined because there is no number in $\{1, 2, 3, 4, 5\}$ for which $5 \times_5 a = 2$.

11. (a) 5; $n +_5 5 = n$ for n in $\{1, 2, 3, 4, 5\}$.

12. (a) Since $5 +_5 5 = 5$, the additive inverse of 5 is 5. Since $2 +_5 3 = 3 +_5 2 = 5$, 2 and 3 are additive inverses. Since $1 +_5 4 = 4 +_5 1 = 5$, 1 and 4 are additive inverses. Therefore, the additive inverses of 1, 2, 3, 4, and 5 are 4, 3, 2, 1, and 5, respectively.

13. (a) Since $2 \times_5 3 = 3 \times_5 2 = 1$, 2 and 3 are multiplicative inverses. Since $1 \times_5 1 = 1$, 1 is its own multiplicative inverse. Since $4 \times_5 4 = 1$, 4 is its own multiplicative inverse. The numbers with multiplicative inverses are 1, 2, 3, and 4.

14. (a) $4 \div_{12} 7 = 4 \times_{12} 7 = 4$

    (c) $3 \div_5 2 = 3 \times_5 3 = 4$

    (e) $2 \div_5 4 = 2 \times_5 4 = 3$

    (g) $4 \times_{12} 7^{-1} = 4 \times_{12} 7 = 4$

(i) $3 \times_5 2^{-1} = 3 \times_5 3 = 4$

(k) $2 \times_5 4^{-1} = 2 \times_5 4 = 3$

15. (c) $(y +_{12} 2) \div_{12} 11 = 3$
$y +_{12} 2 = 3 \times_{12} 11$
$y +_{12} 2 = 9$
$y = 9 +_{12} 10$
$y = 7$
Check: $(7 +_{12} 2) \div_{12} 11 = 9 \div_{12} 11 = 3$.
This is correct because 3 is the only value a in T for which $11 \times_{12} a = 9$.

16. (a) Add the digits: $9 + 9 + 1 + 6 + 4 + 3 + 1 + 1 + 3 = 37$.
We must add a check digit of 3 to obtain a multiple of 10, so we find the code for 99164-3313 3:

17. (a)

Check the sum of the digits: $1 + 0 + 5 + 7 + 4 + 8 + 6 + 5 + 3 + 7 = 46$.
This is not a multiple of 10, so the code is incorrect.

20. (a) $3^3 = 27 = 2 \cdot 12 + 3$; in clock arithmetic $3^3 = 3$.

(c) In clock arithmetic, $3^3 \times_{12} 3^4 = 3 \times_{12} 9 = 3$.

(e) $4^2 = 16 = 1 \cdot 12 + 4$; in clock arithmetic $4^2 = 4$.

(g) $4^{2 \cdot 5} = 4^{10} = 1,048,576 = 87,381 \cdot 12 + 4$; in clock arithmetic $4^{2 \cdot 5} = 4$.

21. (a)
```
     0   R 1   5^16 = 1
   2)1   R 0   5^8 = 1
   2)2   R 1   5^4 = 1
   2)5   R 0   5^2 = 1
   2)10  R 1    5 = 5
   2)21        12-hour clock arithmetic
```

$5^{21} = 5^{1+4+16} = 5^1 \times_{12} 5^4 \times_{12} 5^{16} = 5 \times_{12} 1 \times_{12} 1 = 5$

25. (a) 1 is not a divisor of zero because the only solution of $1 \times_{12} t = 12$ is $t = 12$.
2 is a divisor of zero because $2 \neq 12$ and the equation $2 \times_{12} t = 12$ has a solution not equal to 12 ($t = 6$).
3 is a divisor of zero because $3 \neq 12$ and the equation $3 \times_{12} t = 12$ has solutions not equal to 12 ($t = 4, 8$).
4 is a divisor of zero because $4 \neq 12$ and the equation $4 \times_{12} t = 12$ has solutions not equal to 12 ($t = 3, 6, 9$).
5 is not a divisor of zero because the only solution of $5 \times_{12} t = 12$ is $t = 12$.
6 is a divisor of zero because $6 \neq 12$ and the equation $6 \times_{12} t = 12$ has solutions not equal to 12 ($t = 2, 4, 6, 8, 10$).
7 is not a divisor of zero because the only solution of $7 \times_{12} t = 12$ is $t = 12$.
8 is a divisor of zero because $8 \neq 12$ and the equation $8 \times_{12} t = 12$ has solutions not equal to 12 ($t = 3, 6, 9$).

9 is a divisor of zero because $9 \ne 12$ and the equation $9 \times_{12} t = 12$ has solutions not equal to 12 ($t = 4, 8$).

10 is a divisor of zero because $10 \ne 12$ and the equation $10 \times_{12} t = 12$ has a solution not equal to 12 ($t = 6$).

11 is not a divisor of zero because the only solution of $11 \times_{12} t = 12$ is $t = 12$.

12 is not a divisor of zero because $12 = 12$ but 12 is the additive identity.

The divisors of zero are 2, 3, 4, 6, 8, 9, and 10.

26. Make an orderly list.

y	$y -_{12} 2$	$y -_{12} 3$	$(y -_{12} 2) \times_{12} (y -_{12} 3)$	Solution?
1	11	10	2	No
2	12	11	12	Yes
3	1	12	12	Yes
4	2	1	2	No
5	3	2	6	No
6	4	3	12	Yes
7	5	4	8	No
8	6	5	6	No
9	7	6	6	No
10	8	7	8	No
11	9	8	12	Yes
12	10	9	6	No

The solutions are $y = 2$, $y = 3$, $y = 6$, and $y = 11$.

27. (a) (i) $10 \cdot 0 + 9 \cdot 7 + 8 \cdot 0 + 7 \cdot 8 + 6 \cdot 0 + 5 \cdot 8 + 4 \cdot 2 + 3 \cdot 2 + 2 \cdot 8 + 1 \cdot 7 = 196$
The result in 11-hour clock arithmetic is not 11 because $11 \nmid 196$. This codes is incorrect.

(ii) $10 \cdot 0 + 9 \cdot 2 + 8 \cdot 0 + 7 \cdot 1 + 6 \cdot 3 + 5 \cdot 0 + 4 \cdot 7 + 3 \cdot 2 + 2 \cdot 2 + 1 \cdot 7 = 88$
The result in 11-hour arithmetic is 11 because $11 | 88$. This code is correct.

28. (a) $2 + 4 + 7 + 6 + 3 + 8 + 1 + 1 + 7 = 39$
To obtain a multiple of 10, we need to add 1. The check digit is 1.

33. (a) $D_{60} = \{1, 2, 3, 4, 5, 6, 10, 12, 15, 20, 30, 60\}$
$D_{150} = \{1, 2, 3, 5, 6, 10, 15, 25, 30, 50, 75, 150\}$
So GCD(60, 150) = 30.

34. (a) Factor tree may vary.

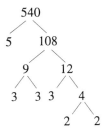

$540 = 2^2 \cdot 3^3 \cdot 5^1$

35. (a)

$$3\overline{)15} \quad 5$$
$$3\overline{)45}$$
$$3\overline{)135}$$
$$2\overline{)270}$$
$$2\overline{)540}$$
$$540 = 2^2 \cdot 3^3 \cdot 5^1$$

$$5\overline{)25} \quad 5$$
$$3\overline{)75}$$
$$2\overline{)150}$$
$$2\overline{)300}$$
$$2\overline{)600}$$
$$600 = 2^3 \cdot 3^1 \cdot 5^2$$

CLASSIC CONUNDRUM   **The Chinese Remainder Problem (page 308)**

Use the constant function of your calculator to find the following arithmetic sequences. Positive integers having remainder 2 when divided by 5:

    2, 7, 12, 17, 22, 27, 32, 37, 42, 47, 52, 57, 62, 67, 72, 77, 82, 87, 92, 97, 102, 107, 112
    117, 122, 127, 132, 137, 142, 147, 152, 157, 162, ...

Positive integers having remainder 3 when divided by 7:

    3, 10, 17, 24, 31, 38, 45, 52, 59, 66, 73, 80, 87, 94, 101, 108, 115, 122, 129, 136, 143,
    150, 157, 164, 171, ...

Positive integers having remainder 4 when divided by 9:

    4, 13, 22, 31, 40, 49, 58, 67, 76, 85, 94, 103, 112, 121, 130, 139, 148, 157, 166, 175, ...

The first number appearing in all three lists is 157. To find the second least integer having this property, note that
$5 = 3^0 \cdot 5^1 \cdot 7^0$, $7 = 3^0 \cdot 5^0 \cdot 7^1$, and $9 = 3^2 \cdot 5^1 \cdot 7^1$.

Therefore, LCM(5, 7, 9) = $3^2 \cdot 5^1 \cdot 7^1 = 315$. Since $157 + \text{LCM}(5, 7, 9) = 157 + 315 = 472$, the second least integer having this property is 472.

Chapter 4 Review Exercises (page 310)

1.

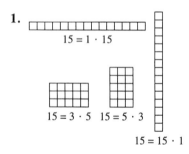

$15 = 1 \cdot 15$

$15 = 3 \cdot 5 \quad 15 = 5 \cdot 3$

$15 = 15 \cdot 1$

2. Answers will vary.

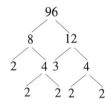

3. (a) $D_{60} = \{1, 2, 3, 4, 5, 6, 10, 12, 15, 20, 30, 60\}$

    (b) $D_{72} = \{1, 2, 3, 4, 6, 8, 9, 12, 18, 24, 36, 72\}$

(c) $D_{60} \cap D_{72} = \{1, 2, 3, 4, 6, 12\}$ so
GCD$\{60, 72\} = 12$

4. (a)
$$\begin{array}{r} 5 \\ 5\overline{)25} \\ 3\overline{)75} \\ 2\overline{)150} \\ 2\overline{)300} \\ 2\overline{)600} \\ 2\overline{)1200} \end{array}$$
$1200 = 2^4 \cdot 3^1 \cdot 5^2$

(b)
$$\begin{array}{r} 7 \\ 7\overline{)49} \\ 5\overline{)245} \\ 3\overline{)735} \\ 2\overline{)1470} \\ 2\overline{)2940} \end{array}$$
$2940 = 2^2 \cdot 3^1 \cdot 5^1 \cdot 7^2$

(c) GCD$(1200, 2940) = 2^2 \cdot 3^1 \cdot 5 \cdot 7^0 = 60$
LCM$(1200, 2940) = 2^4 \cdot 3^1 \cdot 5^2 \cdot 7^2$
$= 58,800$

5. The number 847 will be crossed out as a multiple of 7. (Note that $847 = 7^1 \cdot 11^2$.) Therefore, 847 is composite.

6. (a) Answers will vary. For example,
$15 = 3 \cdot 5$; $5 > \sqrt{15}$.

(b) Yes. $3 \le \sqrt{15}$.

7. Answers will vary. r and s must have a common factor other than 1. For example, $8|16$ and $4|16$, but $32 \nmid 16$.

8. Let $n = 3 \cdot 5 \cdot 7 + 11 \cdot 13 \cdot 17$. Then none of 3, 5, 7, 11, 13, and 17 can divide n (by the result from Exercise 20 of Problem Set 4.1). Therefore, n is prime itself or it has prime factors other than 3, 5, 7, 11, 13 and 17. Indeed, $n = 2536 = 2^3 \cdot 317$, and 317 is prime (since $\sqrt{317} \doteq 17.8$, 317 is odd, and none of the numbers 3, 5, 7, 11, 13, or 17 can divide 317, by the argument presented above). Thus, the number n can be used to generate the primes 2 and 317.

9. (a) 9310 is divisible by 2 and 5:
Divisible by 2 because the last digit is even. Not divisible by 3 because $9 + 3 + 1 + 0 = 13$, $3 \nmid 13$. Divisible by 5 because the last digit is 0 or 5. Not divisible by 11 because $(9 + 1) - (3 + 0) = 7$, and $11 \nmid 7$.

(b) 2079 is divisible by 3 and 11:
Not divisible by 2 because the last digit is odd. Divisible by 3 because $2 + 0 + 7 + 9 = 18$, $3|18$. Not divisible by 5 because the last digit is not 0 or 5. Divisible by 11 because $(2 + 7) - (0 + 9) = 0$, and $11|0$.

(c) 5635 is divisible by 5:
Not divisible by 2 because the last digit is odd. Not divisible by 3 because $5 + 6 + 3 + 5 = 19$, and $3 \nmid 19$. Divisible by 5 because the last digit is 0 or 5. Not divisible by 11 because $(6 + 5) - (5 + 3) = 3$, and $11 \nmid 3$.

(d) 5665 is divisible by 5 and 11:
Not divisible by 2 because the last digit is odd. Not divisible by 3 because $5 + 6 + 6 + 5 = 22$, and $3 \nmid 22$. Divisible by 5 because the last digit is 0 or 5. Divisible by 11 because $(5 + 6) - (6 + 5) = 0$, and $11|0$.

10. (a) 10,197 is divisible by 11.
Use the combined test: $197 - 10 = 187$.
$7 \nmid 187$ because $187 = 26 \cdot 7 + 5$.
$11|187$ because $(1 + 7) - 8 = 0$, and $11|0$.
$13 \nmid 187$ because $187 = 14 \cdot 13 + 5$.

(b) 9373 is divisible by 7 and 13.
Use the combined test: $373 - 9 = 364$.
$7|364$ because $364 = 52 \cdot 7$.
$11 \nmid 364$ because $(3 + 4) - 6 = 1$, and $11 \nmid 1$.
$13|364$ because $274 = 28 \cdot 13$.

(c) 36,751 is divisible by 11 and 13.
Use the combined test: $751 - 36 = 715$.
$7 \nmid 715$ because $715 = 102 \cdot 7 + 1$.
$11|715$ because $(7 + 5) - 1 = 11$, and $11|11$.
$13|715$ because $715 = 55 \cdot 13$.

11. (a) False, because $15 = 3 \cdot 5$, and $3 \nmid 9310$.

(b) True, because $33 = 3 \cdot 11$, $3|2079$, and $11|2079$.

(c) False, because $55 = 5 \cdot 11$, and $11 \nmid 5635$.

(d) True, because $55 = 5 \cdot 11$, $5|5665$, and $11|5665$.

12. (a) The exponents in the prime power representation are 4 and 2, and $(4 + 1)(2 + 1) = 15$. There are 15 divisors.

    (b) Make an orderly list:
    $3^0 \cdot 7^0 = 1$
    $3^0 \cdot 7^1 = 7$
    $3^0 \cdot 7^2 = 49$
    $3^1 \cdot 7^0 = 3$
    $3^1 \cdot 7^1 = 21$
    $3^1 \cdot 7^2 = 147$
    $3^2 \cdot 7^0 = 9$
    $3^2 \cdot 7^1 = 63$
    $3^2 \cdot 7^2 = 441$
    $3^3 \cdot 7^0 = 27$
    $3^3 \cdot 7^1 = 189$
    $3^3 \cdot 7^2 = 1323$
    $3^4 \cdot 7^0 = 81$
    $3^4 \cdot 7^1 = 567$
    $3^4 \cdot 7^2 = 3969$
    In order, the divisors are:
    1, 3, 7, 9, 21, 27, 49, 63, 81, 147, 189, 441, 567, 1323, and 3969.

13. The sum of the odd digits is $2 + 6 + 3 + 1 + d = 12 + d$.
    The sum of the even digits is $7 + 5 + 0 + 2 + 3 = 17$.
    We require $11 | [(12 + d) - 17]$ or $11 | [17 - (12 + d)]$, which is true if $12 + d = 17$. Therefore, $d = 5$.

14. (a) $q = 2^3 \cdot 3^5 \cdot 7^2 \cdot 11^1 \cdot 13^0$
    $m = 2^1 \cdot 3^0 \cdot 7^3 \cdot 11^3 \cdot 13^1$
    $\text{LCM}(q, m) = 2^3 \cdot 3^5 \cdot 7^3 \cdot 11^3 \cdot 13^1$
    $= 11,537,501,976$

    (b) Divide q by its smallest divisor, 2:
    $\frac{q}{2} = \frac{2^3 \cdot 3^5 \cdot 7^2 \cdot 11^1}{2} = 2^2 \cdot 3^5 \cdot 7^2 \cdot 11^1$
    $= 523,908$

15. (a) Cut off a 56 by 56 square, then a 28 by 28 square. The remaining square is 28 by 28.

    (b) GCD(56, 84) = 28

16. (a) From Exercise 15, GCD(56, 84) = 28. In terms of 28 by 28 squares, the 56 by 84 rectangle is 2 squares high and 3 squares wide. Thus, the desired large square is 2 rectangles wide and 3 rectangles high, as shown.

    (b) LCM(56, 84) = 168

17. (a) $D_{63} = \{1, 3, 7, 9, 21, 63\}$
    $D_{91} = \{1, 7, 13, 91\}$
    $D_{63} \cap D_{91} = \{1, 7\}$ so GCD(91, 63) = 7.

    (b) $M_{63} = \{0, 63, 126, 189, 252, 315, 378, 441, 504, 567, 630, 693, 756, 819, 882, 945, 1008, \cdots\}$
    $M_{91} = \{0, 91, 182, 273, 364, 455, 546, 637, 728, 819, 910, \cdots\}$
    $M_{63} \cap M_{91} = \{0, 819, 1638, \cdots\}$, so LCM (63, 91) = 819.

    (c) $7 \cdot 819 = 5733 = 63 \cdot 91$

18. $r = 2^1 \cdot 3^2 \cdot 5^1 \cdot 7^0 \cdot 11^3$
    $s = 2^2 \cdot 3^0 \cdot 5^2 \cdot 7^0 \cdot 11^2$
    $t = 2^3 \cdot 3^1 \cdot 5^0 \cdot 7^1 \cdot 11^3$

(a) $GCD(r, s, t) = 2^1 \cdot 3^0 \cdot 5^0 \cdot 7^0 \cdot 11^2 = 242$

(b) $LCM(r, s, t) = 2^3 \cdot 3^2 \cdot 5^2 \cdot 7^1 \cdot 11^3 = 16,770,600$

19. (a)
$$12{,}100 \overline{)119{,}790} \quad 9 \text{ R } 10{,}890$$
$$10{,}890 \overline{)12{,}100} \quad 1 \text{ R } 1210$$
$$1210 \overline{)10{,}890} \quad 9 \text{ R } 0$$

Thus, $GCD(119{,}790, 12{,}100) = 1210$

(b) $LCM(119{,}790, 12{,}100) = \dfrac{119{,}790 \cdot 12{,}100}{1210} = 1{,}197{,}900$

20. In 2192, because $LCM(17, 13) = 17 \cdot 13 = 221$, and $1971 + 221 = 2192$.

21. (a) $4 + 9 = 13 = 1 \cdot 12 + 1$, so $4 +_{12} 9 = 1$.

(b) $9 -_{12} 4 = 9 +_{12} 8 = 5$

(c) $4 \times 9 = 36 = 3 \cdot 12 + 0$, so $4 \times_{12} 9 = 12$.

(d) $4 \div_{12} 9$ is undefined, because $9 \times_{12} a = 4$ has no solution in T.

(e) $9 + 8 = 17 = 1 \cdot 12 + 5$, so $9 +_{12} 8 = 5$.

(f) $9 \times 12 = 108 = 9 \cdot 12 + 0$, so $9 \times_{12} 12 = 12$.

(g) $4 \div_{12} 12$ is undefined, because $12 \times_{12} a = 4$ has no solution T.

(h) 3 is the only solution in T of $7 \times_{12} t = 9$. Therefore, $9 \div_{12} 7 = 3$.

(i) $9 \times 7 = 63 = 5 \cdot 12 + 3$, so $9 \times_{12} 7 = 3$.

22. (a) $5 + 6 = 11 = 1 \cdot 7 + 4$, so $5 +_7 6 = 4$.

(b) $6 -_7 5 = 6 +_7 2 = 1$

(c) $6 \times 5 = 30 = 4 \cdot 7 + 2$, so $6 \times_7 5 = 2$.

(d) $6 \div_7 5 = 4$, because 4 is the only solution in $\{1, 2, 3, 4, 5, 6, 7\}$ of $5 \times_7 t = 6$.

23. 2, 4, 5, 6, 8, and 10. These are the numbers in $\{1, 2, 3, 4, 5, 6, 7, 8, 9, 10\}$ that have a common factor (other than 1) with 10.

24. (a) Add the digits: $8 + 7 + 2 + 4 + 3 + 1 + 7 + 7 + 2 = 41$.
We need to add 9 to obtain a multiple of 10. The check digit is 9.

(b) Add the digits: $2 + 2 + 0 + 0 + 1 + 8 + 9 + 4 + 1 = 27$.
We need to add 3 to obtain a multiple of 10. The check digit is 3.

25. (a) 8 0 3 2 1 1 5 8 9 3

Check the sum of the digits: $8 + 0 + 3 + 2 + 1 + 1 + 5 + 8 + 9 + 3 = 40$.
This is a multiple of 10, so the bar code appears to be correct. The zip code is 80321 – 1589.

(b)

```
  6   0   6   4   8   9   9   6   0   2
```

Check the sum of the digits: $6 + 0 + 6 + 4 + 8 + 9 + 9 + 6 + 0 + 2 = 50$.
This is a multiple of 10, so the bar code appears to be correct. The zip code is 60648-9960.

26. (a)

$$
\begin{array}{ll}
& 0 \quad R\,1 \quad 3^{32}=8 \\
2\overline{)1} & R\,0 \quad 3^{16}=13 \\
2\overline{)2} & R\,1 \quad 3^{8}=6 \\
2\overline{)5} & R\,1 \quad 3^{4}=12 \\
2\overline{)11} & R\,0 \quad 3^{2}=9 \\
2\overline{)22} & R\,1 \quad 3^{1}=3 \\
2\overline{)45} & \text{23-hour clock arithmetic}
\end{array}
$$

$$3^{45} = 3^{1+4+8+32}$$
$$= 3^1 \times_{23} 3^4 \times_{23} 3^8 \times_{23} 3^{32}$$
$$= 3 \times_{23} 12 \times_{23} 6 \times_{23} 8 = 3$$

(b)

$$
\begin{array}{ll}
& 0 \quad R\,1 \quad 21^{128}=13 \\
2\overline{)1} & R\,1 \quad 21^{64}=6 \\
2\overline{)3} & R\,0 \quad 21^{32}=12 \\
2\overline{)6} & R\,0 \quad 21^{16}=9 \\
2\overline{)12} & R\,1 \quad 21^{8}=3 \\
2\overline{)25} & R\,0 \quad 21^{4}=16 \\
2\overline{)50} & R\,0 \quad 21^{2}=4 \\
2\overline{)100} & R\,0 \quad 21^{1}=21 \\
2\overline{)200} & \text{23-hour clock arithmetic}
\end{array}
$$

$$21^{200} = 21^{8+64+128}$$
$$= 21^8 \times_{23} 21^{64} \times_{23} 21^{128}$$
$$= 3 \times_{23} 6 \times_{23} 13 = 4$$

(c)

$$
\begin{array}{ll}
& 0 \quad R\,1 \quad 5^{128}=6 \\
2\overline{)1} & R\,0 \quad 5^{64}=12 \\
2\overline{)2} & R\,1 \quad 5^{32}=9 \\
2\overline{)5} & R\,1 \quad 5^{16}=3 \\
2\overline{)11} & R\,0 \quad 5^{8}=16 \\
2\overline{)22} & R\,1 \quad 5^{4}=4 \\
2\overline{)45} & R\,0 \quad 5^{2}=2 \\
2\overline{)90} & R\,1 \quad 5^{1}=5 \\
2\overline{)181} &
\end{array}
$$

$$5^{181} = 5^{1+4+16+32+128}$$
$$= 5^1 \times_{23} 5^4 \times_{23} 5^{16} \times_{23} 5^{32} \times_{23} 5^{128}$$
$$= 5 \times_{23} 4 \times_{23} 3 \times_{23} 9 \times_{23} 6$$
$$= 20$$

27. I SURRENDER
Convert to numerical plaintext:
9 19 21 18 18   5 14 4 5 18
Use $C = P +_{26} 7$:
16 26 2 25 25   12 21 11 12 25
Convert to alphabetic ciphertext:
PZBYY LUKLY

28. DRVIO OJWZA MZZST
Since there are many Z's, we might guess (perhaps after some trial and error) that a Z in the ciphertext corresponds to an E in the plaintext. Since Z is the twenty-sixth letter and E is the fifth letter in the alphabet, we guess that $k = 21$. Convert to numerical ciphertext:
4 18 22 9 15   15 10 23 26 1   13 26 26 19 20
Use $P = C +_{26} (26 - k) = C +_{26} 5$:
9 23 1 14 20   20 15 2 5 6   18 5 5 24 25
Convert to alphabetic text:
IWANT TOBEF REEXY
The message was originally encoded using $k = 21$. The message is:
I WANT TO BE FREE XY.

Chapter 4 Test (page 311)

1. (a) False. For example $3|12$ and $6|12$, but $18 \nmid 12$.

   (b) True. If $s = ra$ and $t = sb$, then $t = (ra)b = r(ab)$, so $r|t$.

   (c) True. If $b = au$ and $c = av$, then $b + c = au + av = a(u + v)$, so $a|(b+c)$.

   (d) False. For example, $2 \nmid 7$ and $2 \nmid 9$, but $2|16$.

2. (a) Answers will vary. One possibility is shown.

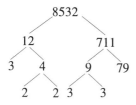

(b) $8532 = 2^2 \cdot 3^3 \cdot 79^1$

(c) Divide 8532 by its smallest divisor (2) to get 4266.

(d) $8532 \cdot 2 = 17,064$

3. (a) 62,418 is divisible by 2 and 3.
Divisible by 2 because the last digit is even.
Divisible by 3 because $6 + 2 + 4 + 1 + 8 = 21$, and $3 | 21$.
Not divisible by 9 because $6 + 2 + 4 + 1 + 8 = 21$, and $9 \nmid 21$.
Not divisible by 11 because $(6 + 4 + 8) - (2 + 1) = 15$, and $11 \nmid 15$.
Not divisible by 33 because $33 = 3 \cdot 11$, and $11 \nmid 62,418$.

(b) 222,789 is divisible by 3.
Not divisible by 2 because the last digit is odd.
Divisible by 3 because $2 + 2 + 2 + 7 + 8 + 9 = 30$, and $3 | 30$.
Not divisible by 9 because $2 + 2 + 2 + 7 + 8 + 9 = 30$, and $9 \nmid 30$.
Not divisible by 11 because $(2 + 7 + 9) - (2 + 2 + 8) = 6$, and $11 \nmid 6$.
Not divisible by 33 because $33 = 3 \cdot 11$, and $11 \nmid 22,789$.

4. (a) $13,534 \overline{)997,476}$ 73 R 9494 $9494 \overline{)13,534}$ 1 R 4040

$4040 \overline{)9494}$ 2 R 1414 $1414 \overline{)4040}$ 2 R 1212

$1212 \overline{)1414}$ 1 R 202 $202 \overline{)1212}$ 6 R 0

GCD(13,534, 997,476) = 202

(b) $\text{LCM}(13,534, \ 997,476) = \dfrac{13,534 \cdot 997,476}{202}$
$= 66,830,892$

5. (a) No. The prime power representation of r contains two 7s, but the prime power representation of m contains only one 7, so $r \nmid m$.

(b) The exponents in the prime power representation of m are 3, 2, 1, and 4, so the number of divisors is $(3 + 1)(2 + 1)(1 + 1)(4 + 1) = 120$.

(c) $m = 2^3 \cdot 5^2 \cdot 7^1 \cdot 11^4$
$n = 2^2 \cdot 5^0 \cdot 7^2 \cdot 11^3$
$\text{GCD}(m, \ n) = 2^2 \cdot 5^0 \cdot 7^1 \cdot 11^3 = 37,268$

(d) $\text{LCM}(m, \ n) = 2^3 \cdot 5^2 \cdot 7^2 \cdot 11^4 = 143,481,800$

6. The divisors of 21 are 1, 3, 7, and 21.

7. **(a)** $7 + 5 = 12 = 1 \cdot 8 + 4$, so $7 +_8 5 = 4$.

   (b) $7 +_{12} 5 = 12 = 1 \cdot 12 + 0$, so $7 +_{12} 5 = 12$.

   (c) 7 is the additive identity in 7-hour clock arithmetic, so $5 -_7 7 = 5$.

   (d) $7 \times 5 = 35 = 4 \cdot 8 + 3$, so $7 \times_8 5 = 3$.

   (e) $5 \times_8 3 = 7$, and there is no other solution to $5 \times_8 t = 7$, so $7 \div_8 5 = 3$.

   (f) $7^5 = 16,807 = 2100 \cdot 8 + 7$, so in 8-hour clock arithmetic, $7^5 = 7$.

8. VICTORY
   Convert to numerical plaintext:
   22 9 3 20 15   18 25
   Use $C = P +_{26} 5$:
   1 14 8 25 20   23 4
   Convert to alphabetic ciphertext:
   ANHYT WD

9. DAZZG ADIZS
   Since there are many Z's, we might guess (perhaps after some trial and error) that a Z in the ciphertext corresponds to an E in the plaintext. Since Z is the twenty-sixth letter and E is the fifth letter in the alphabet, we guess that $k = 21$. Convert to numerical ciphertext:
   4 1 26 26 7   1 4 9 26 19
   Use $P = C +_{26} (26 - k) = C +_{26} 5$:
   9 6 5 5 12   6 9 14 5 24
   Convert to alphabetic text:
   IFEEL FINEX
   The message was originally encoded using $k = 21$. The message is:
   I FEEL FINE XY.

70 *Chapter 4:* Mathematical Reasoning for Elementary Teachers

Chapter 5

Problem Set 5.1 (page 322)

1. (a) The drop must have 5 more black counters than red counters. Two possibilities are ![] and ![].

   (c) The drop must contain the same number of red and black counters or no counters at all. Two possibilities are ● ○ and ● ○ / ○ ●.

2. (a) The drop must contain 3 more black counters than red counters. To use the least number of counters, use 3 black counters only: ● ● ●.

   (c) The drop must contain the same number of red and black counters or no counters at all. To use the least number of counters, use no counters at all: no counters.

3. (a) The drop must contain 5 more black counters than red counters. One possibility is ● ● ● ● ●.

   (c) opp 17

4. (a) At mail time, you are delivered a check for $14. What happens to your net worth?

6. (a) Since the bill is for $15 more than the check you are poorer by $15. The integer is –15.

7. (a) ⟵─┼─┼─┼─●─┼─┼─┼─┼─┼─┼─┼─⟶
       −4−3−2−1 0 1 2 3 4 5 6 7 8

   (b) ⟵─┼─┼─┼─┼─┼─┼─┼─┼─●─┼─┼─┼─⟶
       −4−3−2−1 0 1 2 3 4 5 6 7 8

   (c) ⟵─●─┼─┼─┼─┼─┼─┼─┼─┼─┼─┼─┼─⟶
       −4−3−2−1 0 1 2 3 4 5 6 7 8

   (d) ⟵─┼─┼─┼─┼─┼─┼─┼─┼─┼─┼─┼─┼─●─⟶
       −4−3−2−1 0 1 2 3 4 5 6 7 8

   (e) $\frac{(4+8)}{2} = \frac{12}{2} = 6$

   ⟵─┼─┼─┼─┼─┼─┼─┼─┼─┼─┼─●─┼─┼─⟶
   −4−3−2−1 0 1 2 3 4 5 6 7 8

8. (a) The arrow points 4 units to the right, so it represents 4.

   (c) The arrows point 6 units to the right, so each arrow represents 6.

9. (a) The arrow must point 7 units to the right. One possibility is shown.

   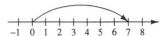

   (c) The arrow must point 9 units to the left. One possibility is shown.

   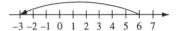

12. (a) Since 34 is 34 units from 0 on a number line, $|34| = 34$.

    (c) Since –76 is 76 units from 0 on a number line, $|-76| = 76$.

13. (a) The equation $|x| = 13$ means that x is 13 units from 0 on a number line. Therefore, $x = 13$ or $x = -13$.

14. (a) To represent –3, the drop needs to have 3 more red counters than black counters. Since there are 3 black counters, we could add red counters until there are 6 red counters—that is, add 4 red counters.

15. (a) 12 red: –12
    11 red + 1 black: –10
    10 red + 2 black: –8
    9 red + 3 black: –6
    8 red + 4 black: –4
    7 red + 5 black: –2
    6 red + 6 black: 0
    5 red + 7 black: 2
    4 red + 8 black: 4
    3 red + 9 black: 6
    2 red + 10 black: 8
    1 red + 11 black: 10
    12 black: 12

17. (a) From Exercise 15, the even integers from –12 to 12 can be represented using all 12 counters, and the odd integers from –11 to 11 can be represented using 11 counters. Furthermore, any number represented by 12 or fewer counters is an integer between –12 and 12

inclusive. So the set of integers that can be represented using 12 or fewer counters is $\{n|n$ is an integer and $-12 \leq n \leq 12\}$.

18. (a) Since there are 2 possibilities (red or black) for each of the 20 counters, there are 2^{20} different appearing rows that can be created. To see this, start by considering easier cares: 1 counter, 2 counters, 3 counters, etc., and look for a pattern. For example, for 4 counters, we have the following arrangements:

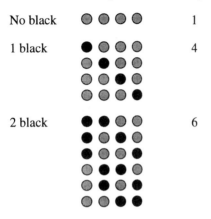

```
No black    ○ ○ ○ ○          1

1 black     ● ○ ○ ○          4
            ○ ● ○ ○
            ○ ○ ● ○
            ○ ○ ○ ●

2 black     ● ● ○ ○          6
            ● ○ ● ○
            ● ○ ○ ●
            ○ ● ● ○
            ○ ● ○ ●
            ○ ○ ● ●

3 black (same as 3 red)      4
4 black (same as 4 red)      1
                            ___
                          16 = 2^4
```

19. (a) Black: $1 + 3 + 5 + 7 + 9 + 11 + 13 + 15 + 17 + 19 = 100$
    Red: $2 + 4 + 6 + 8 + 10 + 12 + 14 + 16 + 18 + 20 = 110$

    To see this consider arrays with 1 row, 2 rows, 3 rows, ⋯, and look for a pattern.

20. Use the theorem on page 255 to find the number of natural number factors.
    (a) $N = (1 + 1)(1 + 1)(1 + 1)(1 + 1)(1 + 1)$
        $= 2^5 = 32$

21. (a) $2^1 \cdot 3^2 \cdot 5^1 = 90$

    (c) $90 \cdot 37{,}800 = 3{,}402{,}000$

22. (a) No. The prime power representation of c contains more 3s than the prime power representation of a.

23. (a)
    $$\begin{array}{r} 7 \\ 5\overline{)35} \\ 5\overline{)175} \\ 2\overline{)350} \\ 2\overline{)700} \\ 2\overline{)1400} \end{array}$$
    $1400 = 2^3 \cdot 5^2 \cdot 7^1$

24. (a)
    $$\begin{array}{r} 1 \text{ R } 891 \\ 4554\overline{)5445} \end{array}$$
    $$\begin{array}{r} 5 \text{ R } 99 \\ 891\overline{)4554} \end{array}$$
    $$\begin{array}{r} 9 \text{ R } 0 \\ 99\overline{)891} \end{array}$$
    $\mathrm{GCD}(4554, 5445) = 99$

JUST FOR FUN Choosing the Right Box (page 332)

Note that there is a symmetry in the problem between apples and oranges. Thus, it probably does not make sense to select the piece of fruit from the box labeled "apples" or the box labeled "oranges." We select from the box labeled "apples and oranges." Since the box is mislabeled, if we obtain an apple it contains only apples and should be so labeled. The remaining boxes contain oranges only and a mixture of apples and oranges. But each box is mislabeled. Thus, the box labeled "oranges" should be relabeled "apples and oranges" and the box labeled "apples and oranges" should be relabeled "oranges." A similar analysis holds if, on our first selection we obtain an orange. Then the boxes should be relabeled as shown.

apples and oranges → oranges
apples → apples and oranges
apples and oranges → apples

JUST FOR FUN Fun with a Flow Chart (page 338)

We start with 27 and x simultaneously and see what results are obtained.

27	x
54	$2x$
58	$2x + 4$
290	$10x + 20$
302	$10x + 32$
3020	$100x + 320$
2700	$100x$
27	x

We obtain the original number whether we start with a 2-digit number or not.

Problem Set 5.2 (page 341)

1. (a)

   $8 + (-3) = 5$

   (c)

   $-8 - (-3) = -5$

   (e)

   $9 + 4 = 13$

   (g)

   $(-9) + 4 = 5$

2. (a) At mail time you receive a bill for $27 and a bill for $13.
   $(-27) + (-13) = -40$

   (c) The mail carrier brings you a check for $27 and a check for $13.
   $27 + 13 = 40$

   (e) At mail time you receive a bill for $41 and a check for $13.
   $(-41) + 13 = -28$

   (g) At mail time you receive a bill for $13 and a check for $41.
   $(-13) + 41 = 28$

3. (a)

   $8 + (-3) = 5$

   (c)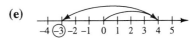

   $(-8) + 3 = -5$

   (e)

   $4 + (-7) = -3$

   (g)

   $(-4) + 7 = 3$

4. (a) $13 - 7 = 13 + (-7)$

   (c) $(-13) - 7 = (-13) + (-7)$

   (e) $3 - 8 = 3 + (-8)$

   (g) $(-8) - 13 = (-8) + (-13)$

5. (a) $27 - (-13) = 27 + 13 = 40$

   (c) $(-13) - 14 = (-13) + (-14)$
   $= -(13 + 14) = -27$

   (e) $(-81) - 54 = (-81) + (-54)$
   $= -(81 + 54) = -135$

   (g) $(-81) + (-54) = -(81 + 54) = -135$

6. (a) $(-41) + 31 = -(41 - 31) = -10$
   The temperature was 10° below zero.

   (b) $(-41) + 31 = -10$

9. (a) $314 - 208 = 106$
   Sam's net worth was more, by $106.

11. (a) $-117 < -24$

    (c) $18 > 12$

    (e) $-5 < 1$

13. Only (a) and (c) are true.

15. No. If $a \geq b$, then it is possible that $a = b$ and so $a > b$ is false.

17. $|x| < 7$ means that, on a number line, x is less than 7 units away from 0. The integers that are less than 7 units away from 0 are –6, –5, –4, –3, –2, –1, 0, 1, 2, 3, 4, 5, and 6.

19. (a) (i) $5 - 11 = 5 + (-11) = -(11 - 5)$
    $= -6$, so $|5 - 11| = |-6| = 6$.

    (iii) $8 - (-7) = 8 + 7 = 15$, so
    $|8 - (-7)| = |15| = 15$.

    (b) (i)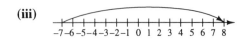

    Distance is 6.

    (iii)

    Distance is 15.

20. (a) (i) $|7 + 2| = |9| = 9$ and
    $|7| + |2| = 7 + 2 = 9$.

    (iii) $|7 + (-6)| = |1| = 1$ and
    $|7| + |-6| = 7 + 6 = 13$.

    (v) $|6 + 0| = |6| = 6$ and
    $|6| + |0| = 6 + 0 = 6$.

22. (a) One way to obtain a magic square using these numbers is to subtract 5 from each of the numbers in the magic square in Exercise 6(a) of Problem Set 1.1. This gives

–1	4	–3
–2	0	2
3	–4	1

Other answers are possible.

23. Answers will vary. One possibility is shown in each case.

    (a) (-1) (0) (1)
        (-2)
        (2)

24. (a) $7 - (-3) = 7 + 3 = 10$ and
    $(-3) - 7 = (-3) + (-7) = -(3 + 7) = -10$

27. (a) $1 - 2 + 3 - \cdots + 99 = 50$

28. (a) ON/AC 3742 + 2167 +/- = , 1575

    (c) ON/AC 2167 +/- − 3742 = , –5909

    (e) ON/AC 3571 − 5624 = +/- , 2053

30. (a) First we write the first 10 Fibonacci numbers:
    $F_1 = 1$, $F_2 = 1$, $F_3 = 2$, $F_4 = 3$, $F_5 = 5$, $F_6 = 8$, $F_7 = 13$, $F_8 = 21$, $F_9 = 34$, $F_{10} = 55$.
    Then we continue the pattern for the sums of the odd-numbered terms of the Fibonacci sequence.
    $$F_1 = 1 \quad = F_2$$
    $$F_1 + F_3 = 1 + 2 = 3 \quad = F_4$$
    $$F_1 + F_3 + F_5 = 1 + 2 + 5 = 8 \quad = F_6$$
    $$F_1 + F_3 + F_5 + F_7 = 1 + 2 + 5 + 13 = 21 \quad = F_8$$
    $$F_1 + F_3 + F_5 + F_7 + F_9 = 1 + 2 + 5 + 13 + 34 = 55 = F_{10}$$

33. (a) Yes. $F_0 + F_1 = 0 + 1 = 1 = F_2$

34. (a) $101 + 3 = 104$

37. (a)

t	h
0	$(-16)\cdot(0)^2 + 96(0) = 0$
1	$(-16)\cdot(1)^2 + 96(1) = 80$
2	$(-16)\cdot(2)^2 + 96(2) = 128$
3	$(-16)\cdot(3)^2 + 96(3) = 144$
4	$(-16)\cdot(4)^2 + 96(4) = 128$
5	$(-16)\cdot(5)^2 + 96(5) = 80$
6	$(-16)\cdot(6)^2 + 96(6) = 0$
7	$(-16)\cdot(7)^2 + 96(7) = -112$

41. (a) Since the last digit, 1, is odd, the number is not divisible by 2. Since the sum of the digits is
$2 + 1 + 4 + 2 + 2 + 1 = 12$, and 12 is divisible by 3, the number is divisible by 3. Since the last digit, 1, is not 0 or 5, the number is not divisible by 5. Since $2 + 4 + 2 = 8$ and $1 + 2 + 1 = 4$, and $8 - 4$ is not divisible by 11, the number is not divisible by 11.
Conclusion: Divisible by 3 only.

42. (a) Using the combined test on page 268, we test the number $965 - 419 = 546$.
Since $546 = 7 \cdot 78$, $7 | 965,419$.
Since $11 \nmid 546$, $11 \nmid 965,419$.
Since $546 = 13 \cdot 42$, $13 | 965,419$.
Conclusion: Divisible by 7 and 13.

JUST FOR FUN   Three on a Bike (page 351)

The problem is that $27 + 2$ should *not* add to 30. Of the $30, $25 is in the till, $2 is in the helper's pocket, and $3 has been returned to the bikers. Alternatively, of the $27, $25 is in the till and $2 is in the helper's pocket.

Problem Set 5.3 (page 356)

1. (a) $7 \cdot 11 = 77$
   (c) $(-7) \cdot 11 = -(7 \cdot 11) = -77$
   (e) $12 \cdot 9 = 108$
   (g) $(-12) \cdot 9 = -(12 \cdot 9) = -108$
   (i) $(-12) \cdot 0 = 0$

2. (a) $36 \div 9 = 4$
   (c) $36 \div (-9) = -(36 \div 9) = -4$
   (e) $(-143) \div 11 = -(143 \div 11) = -13$
   (g) $(-144) \div (-9) = 144 \div 9 = 16$
   (i) $72 \div (21 - 19) = 72 \div 2 = 36$

3. Multiplication: $(-25,753) \cdot (-11) = 283,283$
Division: $283,283 \div (-11) = -25,753$;
$283,283 \div (-25,753) = -11$

5. (a)

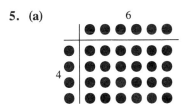

$4 \cdot 6 = 24$

7.

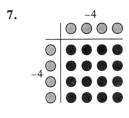

$(-4) \cdot (-4) = 16$

9. (a) Richer by $78; $6 \cdot 13 = 78$

10. (a) $6 \cdot 3 = 18$

12. (a) $3 \cdot (-1) = -3$
$3 \cdot (-2) = -6$
$3 \cdot (-3) = -9$

14. (a) Multiplicative property of zero;
Distributive property of multiplication over addition;
Definition of additive inverse

15. (a) Complete the small circles by noting that $5 - 7 = -2$ and $8 - (-2) = 10$. Then complete the remaining large circle by noting that $7 + 10 = 17$.

(c) Use the method discussed on pages 21–22 in the text. We need to find x, y, and z such that $x + y = -4$, $y + z = 1$, and $x + z = -3$. Since

$$x + y + z = \frac{[(-4) + 1 + (-3)]}{2} = \frac{(-6)}{2} = -3 \text{ and } x + y + z = x + (y + z) = x + 1, \text{ we know that}$$
$x + 1 = -3$, or $x = -4$. Therefore, $y = 0$ and $z = 1$.

16. (b) Complete the upper left circles by noting that $5 - 8 = -3$ and $(-3) + 5 = 2$. The remaining two small circles may be completed with any two numbers whose sum is 2, so the answer is not unique. One possibility is shown.

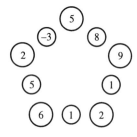

21. (a) $7

(b) $(-105) \div 15 = -7$. A loss of $105, shared among 15 people results in each person losing $7.

24. (b) [ON/AC] 57 [+] [(] 165 [÷] 11 [×OM] [)] [+] 17 [=], 59

Note: Parentheses are unnecessary on a calculator with algebraic logic.

25. (a)(b) The sequence of entries

[ON/AC] 1 [x'] [M+] [=] [-] 2 [x'] [M+] [=] [+] 3 [x'] [M+] [=] ...

successfully computes $s_1, s_2, s_3, \ldots$ and $a_1, a_2, a_3, \ldots$. After each [=] entry, the calculator will show the appropriate entry in the a-sequence and will have the appropriate entry in the s-sequence in memory. To see the s entry press [×OM] right after [=]. Then, to get on with the calculation, press [×OM] again and continue as before. We thus compute the a_n and s_n entries in the following table. The $3s_n \div a_n$ entries are computed separately afterward.

n	a_n	s_n	$3s_n \div a_n$
1	1	1	3
2	-3	5	-5
3	6	14	7
4	-10	30	-9
5	15	55	11
6	-21	91	-13

27. (b) Key in 5 [×] 2 and then press the [=] key 19 times to obtain 2,621,440.

28. (a) $a_3 = 2a_2 + a_1 = 2 \cdot 2 + 1 = 5$
$a_4 = 2a_3 + a_2 = 2 \cdot 5 + 2 = 12$
$a_5 = 2a_4 + a_3 = 2 \cdot 12 + 5 = 29$
$a_6 = 2a_5 + a_4 = 2 \cdot 29 + 12 = 70$

CLASSIC CONUNDRUM   On Stealing Apples
(page 364)

Working backwards, before passing the various watchmen the thief had $2(1 + 2) = 6$, $2(6 + 2) = 16$ and $2(16 + 2) = 36$ apples. He originally stole 36 apples.

Chapter 5 Review Exercises (page 366)

1. (a) $7 - 8 = -1$

   (b) 10 are black and 5 are red, so the number is $10 - 5 = 5$.

   (c) Using the result from Problem Set 5.1, Exercise 16(a), the possible numbers are $-15, -13, -11, ..., 11, 13, 15$.

2. (a) Richer by $12; 12

   (b) Poorer by $37; –37

3. (a) 12

   (b) –24

4. (a) Answers will vary. Any "drop" that shows 5 more red counters than black counters represents the integer –5. For example:
   5 red only;
   1 black + 6 red;
   2 black + 7 red;
   3 black + 8 red;
   4 black + 9 red; etc.

   (b) Any "drop" that shows 6 more black counters than red counters represents the integer 6. For example:
   6 black only;
   7 black + 1 red;
   8 black + 2 red;
   9 black + 3 red;
   10 black + 4 red; etc.

5. Answers will vary.

   (a) At mail time you receive a bill for $114 and a check for $29. $(-114) + 29 = -85$

   (b) The mail carrier brings you a bill for $19 and a check for $66. $(-19) + 66 = 47$

6. (a) The additive inverse of 44 is –44.

   (b) The additive inverse of –61 is $-(-61) = 61$.

7. The drop on the left represents $5 - 3 = 2$ and the drop on the right represents $1 - 5 = -4$, so the diagram represents $2 + (-4) = -2$.

8. The entire drop represents $6 - 7 = -1$ and the counters that are removed represent $1 - 4 = -3$, so the diagram represents $-1 - (-3) = 2$.

9. (a) $45 + (-68) = -(68 - 45) = -23$.
   You are poorer by $23.

   (b) $45 - (-68) = 45 + 68 = 113$.
   You are richer by $113.

10. (a) $5 + (-7) = -(7 - 5) = -2$

    (b) $(-27) - (-5) = (-27) + 5 = -(27 - 5) = -22$

    (c) $(-27) + (-5) = -(27 + 5) = -32$

    (d) $5 - (-7) = 5 + 7 = 12$

    (e) $8 - (-12) = 8 + 12 = 20$

    (f) $8 - 12 = -(12 - 8) = -4$

11. (a) $(-15) - 12 = -(15 + 12) = -27$.
    It is 27° below zero.

    (b) $(-15) - 12 = -27$

12. (a) $(-12) + 37 = 37 - 12 = 25$.
    Her balance is $25.

    (b) $(-12) + 37 = 25$

13. (a) $2 \cdot [(-2) + 4] = 2 \cdot 2 = 4$

    (b) $(-3) \cdot (-5) = 15$

14. (a) The diagram must have 7 rows:

    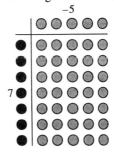

    $7 \cdot (-5) = -35$

    (b) Note that this method is somewhat different from the usual method of representing division because the divisor is on the top of the diagram instead of on the left.

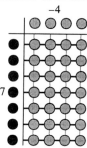

$(-28) \div (-4) = 7$
28 red counters

15. (a)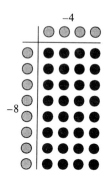

$(-8) \cdot (-4) = 32$

(b) Using the method of Problem 14 we obtain the figure below.

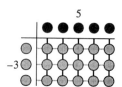

$(-15) \div 5 = -3$
15 red counters

16. (a)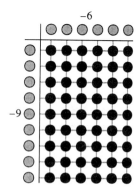

$54 \div (-9) = -6$
54 black counters

(b)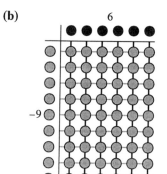

$(-54) \div (-9) = 6$
54 red counters

17. (a) $(-8) \cdot (-7) = 8 \cdot 7 = 56$

(b) $8 \cdot (-7) = -(8 \cdot 7) = -56$

(c) $(-8) \cdot 7 = -(8 \cdot 7) = -56$

(d) $84 \div (-12) = -(84 \div 12) = -7$

(e) $(-84) \div 7 = -(84 \div 7) = -12$

(f) $(-84) \div (-7) = 84 \div 7 = 12$

18. (a) At mail time you receive 7 checks, each for $12.

(b) The mail carrier takes away 7 checks, each for $13.

(c) The mail carrier takes away 7 bills, each for $13.

19. (a)
$$\begin{array}{r} 17 \\ 5\overline{)85} \\ 3\overline{)255} \end{array} \quad \begin{array}{r} 13 \\ 3\overline{)39} \end{array}$$
Since $255 = 3 \cdot 5 \cdot 17$, $39 = 3 \cdot 13$, and 3, 5, 13, and 17 are primes, GCD(255, −39) = 3.

(b)
$$\begin{array}{r} 13 \\ 11\overline{)143} \\ 7\overline{)1001} \end{array} \quad \begin{array}{r} 241 \\ 11\overline{)2651} \end{array}$$
Since $-1001 = (-1) \cdot 7 \cdot 11 \cdot 13$, $2651 = 11 \cdot 241$, and 7, 11, 13, and 241 are primes, GCD(−1001, 2651) = 11.

20. By the division algorithm, n must be of one of these forms: $6q$, $6q + 1$, $6q + 2$, $6q + 3$, $6q + 4$, or $6q + 5$. If n is not divisible by 2, however, then n cannot be of any of the forms $6q$, $6q + 2$ or $6q + 4$. Likewise, if n is not divisible by 3, n cannot be of the form $6q$ or

$6q + 3$. Thus, there must be an integer q such that either $n = 6q + 1$ or $n = 6q + 5$.

Case 1 $n = 6q + 1$

$$n^2 - 1 = (6q+1)^2 - 1$$
$$= (36q^2 + 12q + 1) - 1$$
$$= 36q^2 + 12q$$
$$= 12q(3q+1)$$

If q is even, then $24 \mid 12q$ and $n^2 - 1$ is divisible by 24. If q is odd, then $3q + 1$ is even, so $24 \mid 12(3q+1)$ and hence $n^2 - 1$ is divisible by 24.

Case 2 $n = 6q + 5$

$$n^2 - 1 = (6q+5)^2 - 1$$
$$= (36q^2 + 60q + 25) - 1$$
$$= 36q^2 + 60q + 24$$
$$= 12(3q^2 + 5q + 2)$$
$$= 12(3q+2)(q+1)$$

If q is even, $3q + 2$ is even. If q is odd, $q + 1$ is even. In either case, it follows that $n^2 - 1$ is divisible by 24.

Chapter 5 Test (page 367)

1.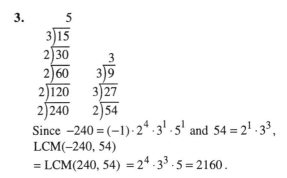

    $(-25) \div (-5) = 5$

2. **(a)** $(-7) + (-19) = -(7 + 19) = -26$

    (b) $(-7) - (-19) = (-7) + 19 = 19 - 7 = 12$

    (c) $7 - (-19) = 7 + 19 = 26$

    (d) $7 + (-19) = -(19 - 7) = -12$

    (e) $(-6859) \div 19 = -(6859 \div 19) = -361$

    (f) $(-24) \cdot 17 = -(24 \cdot 17) = -408$

    (g) $36 \cdot (-24) = -(36 \cdot 24) = -864$

    (h) $(-1155) \div (-11) = 1155 \div 11 = 105$

    (i) $0 \div (-27) = -(0 \div 27) = -0 = 0$

3.
    $$\begin{array}{r} 5 \\ 3\overline{)15} \\ 2\overline{)30} \\ 2\overline{)60} \\ 2\overline{)120} \\ 2\overline{)240} \end{array} \qquad \begin{array}{r} 3 \\ 3\overline{)9} \\ 3\overline{)27} \\ 2\overline{)54} \end{array}$$

    Since $-240 = (-1) \cdot 2^4 \cdot 3^1 \cdot 5^1$ and $54 = 2^1 \cdot 3^3$,
    LCM$(-240, 54)$
    $= $ LCM$(240, 54) = 2^4 \cdot 3^3 \cdot 5 = 2160$.

4. Richer by \$135; $(-5) \cdot (-27) = 135$

5. **(a)**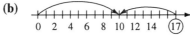

    $(-7) + 10 = 3$

    (b)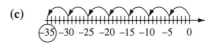

    $10 - (-7) = 17$

    (c)

    $7 \cdot (-5) = -35$

6. Answers may vary.

7. \$381; $129 + 341 - 13 - 47 - 29 = 381$

8. Poorer by \$9; $(-27) \div 3 = -9$

9. **(a)** $(-5) + (-3) = -8$; $(-3) + (-8) = (-11)$;
    $(-8) + (-11) = -19$;
    $(-11) + (-19) = -30$. The sequence is -5, -3, -8, -11, -19, -30.

    (b) $2 - 7 = -5$; $(-5) + 2 = -3$;
    $2 + (-3) = -1$. $(-3) + (-1) = -4$.
    The sequence is $7, -5, 2, -3, -1, -4$.

    (c) Use Guess and Check, make an orderly list and look for a pattern. Using the Fibonacci rule starting with a guess of 1 we obtain
         6, 1, 7, 8, 15, 23.
    Since 15 is too large, we next guess 0 to obtain
         6, 0, 6 6, 12, 18.

Since 12 is still too large, we next guess −1 to obtain
$$6, -1, 5, 4, 9, 13.$$
Notice that 15, 12, and 9 get 3 smaller each time and guess that this pattern will continue. Since we want to arrive at −12 we guess that we should jump 7 more steps. Thus, we guess −8 and obtain
$$6, -8, -2, -10, -12, -22$$
are desired.

10. **(a)** First, note that we can obtain any of the numbers from 1 to 18.
$1 = 1$
$2 = 2$
$1 + 2 = 3$
$5 + (-1) = 4$
$5 = 5$
$5 + 1 = 6$
$5 + 2 = 7$
$5 + 1 + 2 = 8$
$10 + (-1) = 9$
$10 = 10$
$10 + 1 = 11$
$10 + 2 = 12$
$10 + 5 + (-2) = 13$
$10 + 5 + (-1) = 14$
$10 + 5 = 15$
$10 + 5 + 1 = 16$
$10 + 5 + 2 = 17$
$10 + 5 + 2 + 1 = 18$

The number 0 may be obtained as an "empty sum." The numbers −1 to −18 may be obtained by using the additive inverse of each addend shown above. Thus we may obtain any of the numbers −18, −17, −16, ..., −3, −2, −1, 0, 1, 2, 3, ..., 16, 17, 18.

(b) Yes, with the given conditions, there is only one way to represent each number.

Chapter 6

JUST FOR FUN   **Suspicious Simplifications (page 378)**

It works, but it's just lucky. For example,
$$\frac{18}{84} \neq \frac{1}{4}, \quad \frac{27}{76} \neq \frac{2}{6},$$
and so on. A few examples that work out only suggest a property or result. Just one counterexample, like either of the two above, definitely shows that the property does not always hold. In fact, except for numbers like
$$\frac{11}{11}, \frac{22}{22}, \frac{33}{33}, \frac{44}{44} \ldots$$
the only other fraction with two digit numerals with this property is $\frac{26}{65}$.

Problem Set 6.1 (page 382)

1. (a) $\frac{1}{6}$

   (c) $\frac{0}{1}$

   (e) $\frac{2}{6}$

2. (a) Shade 1 of 8 equal subregions. One way to do this is shown.

   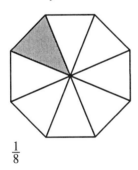

   $\frac{1}{8}$

   (c) Shade 3 of 4 equal subregions. One way to do this is shown.

   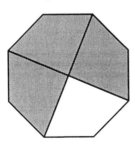

   $\frac{3}{4}$

3. (a) Each tick mark represents $\frac{1}{4}$ of one unit.
   $A: \frac{1}{4}$, $B: \frac{3}{4}$, $C: \frac{6}{4}$ or $\frac{3}{2}$

   (c) Each tick mark represents $\frac{1}{5}$ of one unit.
   $G: \frac{0}{5}$ or 0, $H: \frac{2}{5}$, $I: \frac{8}{5}$

   (e) Each tick mark represents $\frac{1}{5}$ of one unit.
   $M: \frac{-3}{5}$, $N: \frac{-5}{5}$ or -1, $P: \frac{-8}{5}$

4. (a) Answers may vary. One method is shown.

5. (a) $\frac{20}{60}$ or $\frac{1}{3}$ (since an hour is 60 minutes)

   (c) $\frac{5}{7}$ (since a week is 7 days)

   (e) $\frac{25}{100}$ or $\frac{1}{4}$ (since a quarter is 25¢ and a dollar is 100¢)

   (g) $\frac{2}{3}$ (since a yard is 3 feet)

7. (a) $\frac{3}{6} = \frac{1}{2}$

9. Answers may vary.

   (a)

   $\frac{2}{4} \quad = \quad \frac{4}{8}$

10. (a) If $\frac{4}{5} = \frac{x}{30}$, then $4 \cdot 30 = 5 \cdot x$, so $x = 120 \div 5 = 24$. The missing integer is 24.

    (c) If $\frac{-7}{25} = \frac{z}{500}$, then $(-7) \cdot 500 = 25 \cdot z$, so $z = (-3500) \div 25 = -140$. The missing integer is -140.

11. (a) A common denominator is LCM(42, 7) = 42. We can write the

fractions as $\frac{18}{42}$ and $\frac{3 \cdot 6}{7 \cdot 6} = \frac{18}{42}$. They are equivalent.

(c) A common denominator is LCM(25, 500) = 500. We can write the fractions as $\frac{9 \cdot 20}{25 \cdot 20} = \frac{180}{500}$ and $\frac{140}{500}$. Since $\frac{180}{500} \neq \frac{140}{500}$, they are not equivalent.

12. (a) Since 78 · 168 = 13,104 and 24 · 546 = 13,104, we conclude that 78 · 168 = 24 · 546 and the fractions are equivalent.

13. (a) Yes. 4 · 3 = 12 and 9 · 3 = 27.

    (c) Yes. 4 + 3 = 7 and 9 + 3 = 12.

14. (a) $\frac{84}{144} = \frac{42}{72} = \frac{21}{36} = \frac{7}{12}$

    (c) $\frac{-930}{1290} = \frac{-93}{129} = \frac{-31}{43}$

15. (a)

    $$\begin{array}{r} 3 \\ 2\overline{)6} \\ 2\overline{)12} \\ 2\overline{)24} \\ 2\overline{)48} \\ 2\overline{)96} \end{array} \quad \begin{array}{r} 3\overline{)9} \\ 2\overline{)18} \\ 2\overline{)36} \\ 2\overline{)72} \\ 2\overline{)144} \\ 2\overline{)288} \end{array}$$

    $\frac{96}{288} = \frac{2^5 \cdot 3^1}{2^5 \cdot 3^2} = \frac{1}{3^1} = \frac{1}{3}$

16. Answers will vary. Possible answers (using the least common denominator) are shown.

    (a) The least common denominator is LCM(11, 5) = 55. We write $\frac{3 \cdot 5}{11 \cdot 5} = \frac{15}{55}$ and $\frac{2 \cdot 11}{5 \cdot 11} = \frac{22}{55}$. The fractions are equivalent to $\frac{15}{55}$ and $\frac{22}{55}$.

    (c) The least common denominator is LCM(3, 8, 6) = 24. We write $\frac{4 \cdot 8}{3 \cdot 8} = \frac{32}{24}$, $\frac{5 \cdot 3}{8 \cdot 3} = \frac{15}{24}$, and $\frac{1 \cdot 4}{6 \cdot 4} = \frac{4}{24}$. The fractions are equivalent to $\frac{32}{24}$, $\frac{15}{24}$, and $\frac{4}{24}$.

17. (a) The least common denominator is LCM(8, 6) = 24. We write $\frac{3 \cdot 3}{8 \cdot 3} = \frac{9}{24}$ and $\frac{5 \cdot 4}{6 \cdot 4} = \frac{20}{24}$. The fractions are equivalent to $\frac{9}{24}$ and $\frac{20}{24}$.

    (c) The least common denominator is LCM(12,32) = 96. We write $\frac{17 \cdot 8}{12 \cdot 8} = \frac{136}{96}$ and $\frac{7 \cdot 3}{32 \cdot 3} = \frac{21}{96}$. The fractions are equivalent to $\frac{136}{96}$ and $\frac{21}{96}$.

18. (a) Since $2 \cdot 12 > 3 \cdot 7$, $\frac{2}{3} > \frac{7}{12}$. The rational numbers in order are $\frac{7}{12}$, $\frac{2}{3}$.

    (c) Since $5 \cdot 36 > 6 \cdot 29$, $\frac{5}{6} > \frac{29}{36}$. The rational numbers in order are $\frac{29}{36}$, $\frac{5}{6}$.

19. (a) True. Given two fractions, two equivalent fractions with a common denominator may be found by finding a common multiple of the two original denominators. Once these fractions, say $\frac{a}{c}$ and $\frac{b}{c}$ are found, infinitely many more pairs of equivalent fractions can be found; namely, $\frac{a \cdot n}{c \cdot n}$ and $\frac{b \cdot n}{c \cdot n}$ for any integer n other than 0.

    (c) False. Given any positive fraction, a smaller positive fraction may be found by multiplying the denominator by 2. So there cannot be a least positive fraction.

20. (a) As in Example 6.1(d), the fractions equivalent to $\frac{3}{5}$ can be written in the form $\frac{3 \cdot n}{5 \cdot n}$, where n is a nonzero integer. The set of these fractions can be written as $\left\{..., \frac{-9}{-15}, \frac{-6}{-10}, \frac{-3}{-5}, \frac{3}{5}, \frac{6}{10}, \frac{9}{15}, ...\right\}$ or $\left\{\frac{a}{b} \,\middle|\, a = 3 \cdot n, \, b = 5 \cdot n \text{ for any integer } n \neq 0\right\}$.

(c) The fractions equivalent to 0 (or $\frac{0}{1}$) can be written as $\frac{0 \cdot n}{1 \cdot n}$, where n is a nonzero integer. The set of these fractions can be written as $\{\ldots \frac{0}{-3}, \frac{0}{-2}, \frac{0}{-1}, \frac{0}{1}, \frac{0}{2}, \frac{0}{3}, \ldots\}$ or $\{\frac{0}{n} \mid n \text{ is any integer}, n \neq 0\}$.

22. (a) $\frac{4}{8}$

(d) $\frac{1}{5}$

(f) 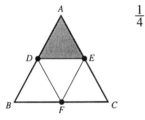 $\frac{1}{4}$

23. (a) Since $\frac{1}{1} = \frac{30}{30}$, $\frac{2}{1} = \frac{60}{30}$, $\frac{3}{2} = \frac{45}{30}$, $\frac{5}{3} = \frac{50}{30}$, and $\frac{8}{5} = \frac{48}{30}$, the fractions in order are $\frac{30}{30}, \frac{45}{30}, \frac{48}{30}, \frac{50}{30}, \frac{60}{30}$, or $\frac{1}{1}, \frac{3}{2}, \frac{8}{5}, \frac{5}{3}, \frac{2}{1}$.

(b) The pattern seems to be that each fraction goes between the previous two, so we predict that $\frac{13}{8}$ will be between $\frac{8}{5}$ and $\frac{5}{3}$. We confirm that $\frac{8}{5} < \frac{13}{8} < \frac{5}{3}$ by noting that $8 \cdot 8 < 5 \cdot 13$ and $13 \cdot 3 < 8 \cdot 5$.

24. (a) The mediant is $\frac{3+4}{4+5} = \frac{7}{9}$; $\frac{3}{4} < \frac{7}{9} < \frac{4}{5}$ since $3 \cdot 9 < 4 \cdot 7$ and $7 \cdot 5 < 9 \cdot 4$.

25. (a) Let $\frac{a}{b} < \frac{c}{d}$ with b and d positive so $ad < bc$. But then $ad + mcd < bc + mcd$ where m is any natural number. Therefore, $(a + mc)d < (b + md)c$ and this implies that
$$\frac{a+mc}{b+md} < \frac{c}{d}.$$
Also, if $ad < bc$, then $mad < mbc$. But then $ab + mad < ab + mbc$ and $a(b + md) < b(a + mc)$. This last statement implies that
$$\frac{a}{b} < \frac{a+mc}{b+md}.$$
So
$$\frac{a}{b} < \frac{a+mc}{b+md} < \frac{c}{d}.$$

Note: This problem can also be done by noting that $\frac{a}{b} < \frac{mc}{md}$, and applying the result from Exercise 24(b).

(b) Given any two distinct rational numbers, $\frac{a}{b}$ and $\frac{c}{d}$, where $\frac{a}{b} < \frac{c}{d}$ and b and d are positive, $\frac{a+mc}{b+md}$ is a rational number between $\frac{a}{b}$ and $\frac{c}{d}$ for any natural number m. Since there are infinitely many choices for m, and each choice gives a distinct rational number, there are infinitely many rational numbers between $\frac{a}{b}$ and $\frac{c}{d}$.

(Indeed, $\frac{a}{b} < \frac{a+c}{b+d} < \frac{a+2c}{b+2d} < \cdots$
$\cdots < \frac{a+mc}{b+md} < \cdots < \frac{c}{d}$.)

26. (a) Let a, b, and c be positive integers, and suppose $a < c$. Then $ab < bc$ and so $\frac{a}{b} < \frac{c}{b}$.

27. (a) Answers will vary. One possibility for each tiling is shown.

(b) Regardless of the answer to part (a), in each case one-third of the rhombuses will have each of the three possible orientations.

28. (a) The fifth Farey sequence is obtained by inserting the numbers $\frac{1}{5}$, $\frac{2}{5}$, $\frac{3}{5}$, and $\frac{4}{5}$ into the appropriate location in the fourth Farey sequence.

$\mathcal{F}_5$: $\frac{0}{1}, \frac{1}{5}, \frac{1}{4}, \frac{1}{3}, \frac{2}{5}, \frac{1}{2}, \frac{3}{5}, \frac{2}{3}, \frac{3}{4}, \frac{4}{5}, \frac{1}{1}$

The sixth Farey sequence is obtained by inserting the numbers $\frac{1}{6}$ and $\frac{5}{6}$ into the appropriate location in $\mathcal{F}_5$. (Note that $\frac{2}{6}$, $\frac{3}{6}$, and $\frac{4}{6}$ are already present in the form $\frac{1}{3}$, $\frac{1}{2}$, and $\frac{2}{3}$.)

$\mathcal{F}_6$: $\frac{0}{1}, \frac{1}{6}, \frac{1}{5}, \frac{1}{4}, \frac{1}{3}, \frac{2}{5}, \frac{1}{2}, \frac{3}{5}, \frac{2}{3}, \frac{3}{4}, \frac{4}{5}, \frac{5}{6}, \frac{1}{1}$

29. (a) Since $\frac{3}{4} = \frac{18}{24}$, the tank contains 18 gallons.

30. Answers may vary.

(a) Since $\frac{310}{498}$ is about $\frac{300}{500}$, a simpler fraction is $\frac{3}{5}$.

(c) Since $\frac{9}{35}$ is about $\frac{9}{36}$, a simpler fraction is $\frac{1}{4}$.

31. Yes. Lakeside won $\frac{19}{25}$ of their games while Shorecrest won $\frac{16}{21}$ of their games, and $\frac{16}{21} > \frac{19}{25}$ because $16 \cdot 25 > 21 \cdot 19$.

32. (a) First half: Carol. $\frac{38}{50} > \frac{30}{40}$ because $38 \cdot 40 > 50 \cdot 30$.
Second half: Carol. $\frac{42}{70} > \frac{14}{24}$ because $42 \cdot 24 > 70 \cdot 14$.

33. (a) Since there are two possible outcomes (heads or tails) and only one of these is "successful," the probability is $\frac{1}{2}$.

(c) Since there are 3 even numbers out of six possible outcomes, the probability is $\frac{3}{6}$ or $\frac{1}{2}$.

35. (a) $2 - (-3) = 5$, $(-3) - 5 = -8$,
$5 - (-8) = 13$, $(-8) - 13 = -21$,
$13 - (-21) = 34$, $(-21) - 34 = -55$,
$34 - (-55) = 89$, $(-55) - 89 = -144$,
$89 - (-144) = 233$, $(-144) - 233 = -377$

36. (a) Answers will vary. One possibility is $m = -4$ and $n = -3$, since
$8 \cdot (-4) - 11 \cdot (-3) = -32 - (-33) = 1$.
Another possibility is $m = 7$ and $n = 5$, since $8 \cdot 7 - 11 \cdot 5 = 56 - 55 = 1$.

37. Divisibility by 3:
$4 + 9 + D + 8 + 4 = 25 + D$.
Since $3 | 25 + D$, we know that $D = 2$, $D = 5$, or $D = 8$.
Divisibility by 8: The last three digits are 284, 584, or 884. Of these, only 584 is divisible by 8. Therefore, $D = 5$.

JUST FOR FUN The Sultan's Estate (page 389)

Since $\left(\frac{1}{2}\right) + \left(\frac{1}{3}\right) + \left(\frac{1}{9}\right) = \frac{17}{18}$, the canny old Sultan did not leave all his horses to his sons. The uncle, realizing this, saw that if he included his horse before the division, the sons could each receive a whole number of horses and the uncle could then take back his own horse.

JUST FOR FUN Bookworm Math (page 392)

The bookworm only needs to eat through the front cover of Volume 1 and the back cover of Volume 2 for a total of $\frac{1}{8}+\frac{1}{8}=\frac{1}{4}$ inch.

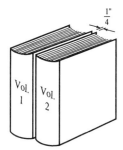

Problem Set 6.2

1. (a) $\frac{1}{3}+\frac{1}{2}=\frac{5}{6}$

2. (a)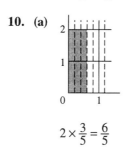
$\frac{2}{5} + \frac{6}{5} = \frac{8}{5}$

3. (a)

$\frac{3}{4}+\frac{-2}{4}=\frac{1}{4}$

4. (a) $\frac{2}{7}+\frac{3}{7}=\frac{2+3}{7}=\frac{5}{7}$

   (c) $\frac{3}{8}+\frac{11}{24}=\frac{9}{24}+\frac{11}{24}=\frac{9+11}{24}=\frac{20}{24}=\frac{5}{6}$

   (e) $\frac{5}{12}+\frac{17}{20}=\frac{25}{60}+\frac{51}{60}=\frac{25+51}{60}=\frac{76}{60}=\frac{19}{15}$

   (g) $\frac{-57}{100}+\frac{13}{10}=\frac{-57}{100}+\frac{130}{100}=\frac{(-57)+130}{100}$
   $=\frac{73}{100}$

5. (a) Since $9 = 2 \cdot 4 + 1$,
   $\frac{9}{4}=\frac{2\cdot 4+1}{4}=2+\frac{1}{4}=2\frac{1}{4}$.

   (c) Since $111 = 4 \cdot 23 + 19$,
   $\frac{111}{23}=\frac{4\cdot 23+19}{23}=4+\frac{19}{23}=4\frac{19}{23}$.

6. (a) $2\frac{3}{8}=\frac{2\cdot 8+3}{8}=\frac{19}{8}$

   (c) $111\frac{2}{5}=\frac{111\cdot 5+2}{5}=\frac{557}{5}$

7. (a) $\frac{5}{6}-\frac{1}{4}=\frac{7}{12}$

8. (a) $\frac{5}{8}-\frac{2}{8}=\frac{5-2}{8}=\frac{3}{8}$

   (c) $2\frac{2}{3}-1\frac{1}{3}=\frac{8}{3}-\frac{4}{3}=\frac{8-4}{3}=\frac{4}{3}$

   (e) $\frac{6}{8}-\frac{5}{12}=\frac{9}{12}-\frac{5}{12}=\frac{9-5}{12}=\frac{4}{12}=\frac{1}{3}$

   (g) $\frac{137}{214}-\frac{-1}{3}=\frac{137\cdot 3-214\cdot(-1)}{214\cdot 3}=\frac{625}{642}$

9. (a) $3\times\frac{5}{2}=\frac{15}{2}$

10. (a)

$2\times\frac{3}{5}=\frac{6}{5}$

12. (a) The reciprocal of $\frac{3}{8}$ is $\frac{8}{3}$.

    (c) The reciprocal of $2\frac{1}{4}=\frac{9}{4}$ is $\frac{4}{9}$.

    (e) The reciprocal of $5=\frac{5}{1}$ is $\frac{1}{5}$.

13. We may assume $b > 0$. If $0 < \frac{a}{b} < 1$, then $0 < a < b$, so $\frac{b}{a} > 1$. If $\frac{c}{d}$ is nearly 1, then c and d are nearly equal, so $\frac{d}{c}$ is also close to 1.

15. (a) $\frac{2}{5}\div\frac{3}{4}-\frac{2}{5}\cdot\frac{4}{3}=\frac{8}{15}$

    (c) $\frac{100}{33}\div\frac{10}{3}=\frac{100}{33}\cdot\frac{3}{10}=\frac{300}{330}=\frac{10}{11}$

    (e) $3\div 5\frac{1}{4}=\frac{3}{1}\div\frac{21}{4}=\frac{3}{1}\cdot\frac{4}{21}=\frac{12}{21}=\frac{4}{7}$

17. (a) $\left[\frac{1}{2} - \left(\frac{2}{3} \cdot \frac{3}{8}\right)\right] \div \left(\frac{2}{7} - \frac{1}{14}\right)$

$= \left(\frac{1}{2} - \frac{6}{24}\right) \div \left(\frac{2}{7} - \frac{1}{14}\right)$

$= \left(\frac{2}{4} - \frac{1}{4}\right) \div \left(\frac{4}{14} - \frac{1}{14}\right)$

$= \frac{1}{4} \div \frac{3}{14} = \frac{1}{4} \cdot \frac{14}{3} = \frac{14}{12} = \frac{7}{6}$

18. (a) To find $\frac{e}{f}$ such that $\frac{2}{3} + \frac{e}{f} = \frac{3}{4}$, we subtract: $\frac{e}{f} = \frac{3}{4} - \frac{2}{3} = \frac{9}{12} - \frac{8}{12} = \frac{1}{12}$. Therefore, $\frac{2}{3} + \frac{1}{12} = \frac{3}{4}$.

19. Yes. $\frac{1}{2} + \frac{1}{3} + \frac{1}{9}$ is the fraction of the horses given to the sons—namely, $\frac{17}{18}$, since 17 of the 18 horses are given to the sons. Check: $\frac{1}{2} + \frac{1}{3} + \frac{1}{9} = \frac{9}{18} + \frac{6}{18} + \frac{2}{18} = \frac{17}{18}$.

20. (a) Top right: Using the top row,
$1 - \frac{1}{2} - \frac{1}{12} = \frac{12}{12} - \frac{6}{12} - \frac{1}{12} = \frac{5}{12}$.
Bottom left: Using the diagonal,
$1 - \frac{5}{12} - \frac{1}{3} = \frac{12}{12} - \frac{5}{12} - \frac{4}{12} = \frac{3}{12} = \frac{1}{4}$.
Bottom middle: Using the middle column,
$1 - \frac{1}{12} - \frac{1}{3} = \frac{12}{12} - \frac{1}{12} - \frac{4}{12} = \frac{7}{12}$.
Bottom right: Using the diagonal,
$1 - \frac{1}{2} - \frac{1}{3} = \frac{12}{12} - \frac{6}{12} - \frac{4}{12} = \frac{2}{12} = \frac{1}{6}$.
Middle left: Using the left column,
$1 - \frac{1}{2} - \frac{1}{4} = \frac{4}{4} - \frac{2}{4} - \frac{1}{4} = \frac{1}{4}$.
Middle right: Using the right column,
$1 - \frac{5}{12} - \frac{1}{6} = \frac{12}{12} - \frac{5}{12} - \frac{2}{12} = \frac{5}{12}$.
Therefore, we obtain

$\frac{1}{2}$	$\frac{1}{12}$	$\frac{5}{12}$
$\frac{1}{4}$	$\frac{1}{3}$	$\frac{5}{12}$
$\frac{1}{4}$	$\frac{7}{12}$	$\frac{1}{6}$

To confirm that this is a Magic Fraction Square, we check the bottom two rows:
$\frac{1}{4} + \frac{1}{3} + \frac{5}{12} = \frac{3}{12} + \frac{4}{12} + \frac{5}{12} = \frac{12}{12} = 1$ and
$\frac{1}{4} + \frac{7}{12} + \frac{1}{6} = \frac{3}{12} + \frac{7}{12} + \frac{2}{12} = \frac{12}{12} = 1$.

21. (a) $S_5 = S_4 + \frac{1}{5} = \frac{25}{12} + \frac{1}{5} = \frac{125}{60} + \frac{12}{60} = \frac{137}{60}$

$S_6 = S_5 + \frac{1}{6} = \frac{137}{60} + \frac{1}{6} = \frac{137}{60} + \frac{10}{60} = \frac{147}{60}$

$= \frac{49}{20}$

$S_7 = S_6 + \frac{1}{7} = \frac{49}{20} + \frac{1}{7} = \frac{49 \cdot 7 + 20 \cdot 1}{20 \cdot 7}$

$= \frac{363}{140}$

$S_8 = S_7 + \frac{1}{8} = \frac{363}{140} + \frac{1}{8} = \frac{726}{280} + \frac{35}{280}$

$= \frac{761}{280}$

22. The mixed number $A\frac{b}{c}$ is equal to $A + \frac{b}{c}$. Rewrite this using a common denominator:

$A + \frac{b}{c} = \frac{A}{1} \times \frac{c}{c} + \frac{b}{c}$

$= \frac{Ac}{c} + \frac{b}{c}$

$= \frac{Ac + b}{c}$.

23. (a) Using the common denominator algorithm, $\frac{7}{12} \div \frac{11}{12} = \frac{7}{11}$.
Using the divide-numerators and denominators algorithm,

$\frac{4}{15} \div \frac{2}{3} = \frac{4 \div 2}{15 \div 3} = \frac{2}{5}$ and

$\frac{19}{210} \div \frac{19}{70} = \frac{19 \div 19}{210 \div 70} = \frac{1}{3}$.

24. (a) $\frac{1}{2} + \frac{1}{3} + \frac{1}{15} + \frac{1}{50}$

$= \frac{75}{150} + \frac{50}{150} + \frac{10}{150} + \frac{3}{150} = \frac{138}{150} = \frac{23}{25}$

25. (a) Following the hint for $\frac{2}{9}$, note that half of $9 + 1$ is 5, so we expect one of the addend fractions to be $\frac{1}{5}$. The other fraction is $\frac{2}{9} - \frac{1}{5} = \frac{10}{45} - \frac{9}{45} = \frac{1}{45}$.
Thus $\frac{2}{9} = \frac{1}{5} + \frac{1}{45}$. Likewise, half of $11 + 1$ is 6, and $\frac{2}{11} - \frac{1}{6} = \frac{12}{66} - \frac{11}{66} = \frac{1}{66}$.
Therefore, $\frac{2}{11} = \frac{1}{6} + \frac{1}{66}$.

26. (a) $a_1 = \frac{1}{2} = 1 - \frac{1}{2}$, $a_2 = \frac{1}{2} + \frac{1}{4} = \frac{3}{4} = 1 - \frac{1}{4}$,

$a_3 = \frac{1}{2} + \frac{1}{4} + \frac{1}{8} = \frac{7}{8} = 1 - \frac{1}{8}$

27. (a) It terminates.

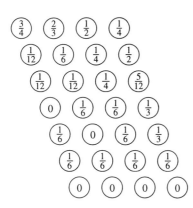

28. (a) The process terminates when one row consists of all ones.

29. In each case, the number of votes must be rounded *up* to the next whole number.

3 members: $\frac{3}{4} \cdot 3 = \frac{9}{4} = 2\frac{1}{4}$, so 3 votes are needed.

4 members: $\frac{3}{4} \cdot 4 = \frac{12}{4} = 3$, so 3 votes are needed.

5 members: $\frac{3}{4} \cdot 5 = \frac{15}{4} = 3\frac{3}{4}$, so 4 votes are needed.

6 members: $\frac{3}{4} \cdot 6 = \frac{18}{4} = 4\frac{1}{2}$, so 5 votes are needed.

7 members: $\frac{3}{4} \cdot 7 = \frac{21}{4} = 5\frac{1}{4}$, so 6 votes are needed.

8 members: $\frac{3}{4} \cdot 8 = \frac{24}{4} = 6$, so 6 votes are needed.

31. Each bow requires $1\frac{1}{2} \div 6 = \frac{3}{2} \cdot \frac{1}{6} = \frac{3}{12} = \frac{1}{4}$ yard of ribbon. Since $5\frac{3}{4} \div \frac{1}{4} = \frac{23}{4} \cdot \frac{4}{1} = 23$, 23 bows can be made.

33. Since $28 was $\frac{2}{3}$ of the original price, the original price was $28 \div \frac{2}{3} = \frac{28}{1} \cdot \frac{3}{2} = \frac{84}{2} = 42$, or $42.00.

37. (a) $\frac{168}{48} = \frac{84}{24} = \frac{42}{12} = \frac{21}{6} = \frac{7}{2}$

39. Since $9 = 3^2$, $12 = 2^2 \cdot 3^1$, and $15 = 3^1 \cdot 5^1$, the least common denominator is $2^2 \cdot 3^2 \cdot 5 = 180$. The sum is $\frac{4}{9} + \frac{5}{12} + \frac{4}{15} = \frac{80}{180} + \frac{75}{180} + \frac{48}{180} = \frac{203}{180}$.

Problem Set 6.3 (page 418)

1. Commutative and associative properties for addition:

$\left(\left(3\frac{1}{5} + 2\frac{2}{5}\right) + 8\frac{1}{5}\right) = (3 + 2 + 8) + \left(\frac{1}{5} + \frac{2}{5} + \frac{1}{5}\right) = 13 + \frac{4}{5} = 13\frac{4}{5}$

3. (a) $\frac{-4}{5}$

(c) $\frac{8}{3}$

4. (a) $\frac{1}{6} + \frac{2}{-3} = \frac{1}{6} + \left(-\frac{2}{3}\right) = \frac{1}{6} - \frac{4}{6} = \frac{-3}{6} = \frac{-1}{2}$

(c) $\frac{9}{4} + \frac{-7}{8} = \frac{18}{8} + \frac{-7}{8} = \frac{11}{8}$
The most useful property here is equivalence of fractions.

5. (a) $\frac{2}{5} - \frac{3}{4} = \frac{2 \cdot 4 - 5 \cdot 3}{5 \cdot 4} = \frac{-7}{20}$

   (c) $\frac{3}{8} - \frac{1}{12} = \frac{9}{24} - \frac{2}{24} = \frac{7}{24}$

   (e) $2\frac{1}{3} - 5\frac{3}{4} = \frac{7}{3} - \frac{23}{4} = \frac{7 \cdot 4 - 3 \cdot 23}{3 \cdot 4} = \frac{-41}{12}$
   $= \frac{-3}{2}$
   The most useful property here is equivalence of fractions.

6. (a) $\frac{3}{5} \cdot \frac{7}{8} \cdot \left(\frac{5}{3}\right) = \frac{3 \cdot 7 \cdot 5}{5 \cdot 8 \cdot 3} = \frac{7}{8}$

   (c) $\frac{-4}{3} \cdot \frac{6}{-16} = \frac{-24}{-48} = \frac{1}{2}$

   (e) $\frac{14}{15} \cdot \frac{60}{7} = \frac{14 \cdot 60}{7 \cdot 15} = \frac{2 \cdot 4}{1} = \frac{8}{1}$ or 8
   The most useful property here is equivalence of fractions.

7. (a) $\frac{2}{3}$

   (c) $\frac{-11}{-4}$ or $\frac{11}{4}$

   (e) $-\frac{1}{2}$

8. (a) $\frac{2}{3} \cdot \frac{4}{7} + \frac{2}{3} \cdot \frac{3}{7} = \frac{2}{3} \cdot \left(\frac{4}{7} + \frac{3}{7}\right) = \frac{2}{3} \cdot 1 = \frac{2}{3}$

   (c) $\frac{4}{7} \cdot \frac{3}{2} - \frac{4}{7} \cdot \frac{6}{4} = \frac{4}{7} \cdot \frac{3}{2} - \frac{4}{7} \cdot \frac{3}{2} = \frac{12}{14} - \frac{12}{14}$
   $= 0$

9. (a) Addition of rational numbers —definition

10. Multiply both sides by $\frac{7}{4}$ to get $\left(\frac{a}{b} \cdot \frac{4}{7}\right) \cdot \frac{7}{4} = \frac{2}{3} \cdot \frac{7}{4}$. By the associative property of multiplication and the multiplicative inverse property, $\left(\frac{a}{b} \cdot \frac{4}{7}\right) \cdot \frac{7}{4} = \frac{a}{b} \cdot \left(\frac{4}{7} \cdot \frac{7}{4}\right) = \frac{a}{b} \cdot 1 = \frac{a}{b}$.

$\frac{a}{b} \cdot \frac{4}{7} = \frac{2}{3}$	Given
$\left(\frac{a}{b} \cdot \frac{4}{7}\right) \cdot \frac{7}{4} = \frac{2}{3} \cdot \frac{7}{4}$	Multiply both sides by $\frac{7}{4}$.
$\left(\frac{a}{b} \cdot \frac{4}{7}\right) \cdot \frac{7}{4} = \frac{7}{6}$	$\frac{2}{3} \cdot \frac{7}{4} = \frac{14}{12} = \frac{7}{6}$
$\frac{a}{b} \cdot \left(\frac{4}{7} \cdot \frac{7}{4}\right) = \frac{7}{6}$	Associative property of multiplication
$\frac{a}{b} \cdot 1 = \frac{7}{6}$	Multiplicative inverse property
$\frac{a}{b} = \frac{7}{6}$	One is a multiplicative identity.

11. (a)
$4x + 3 = 0$	Given
$4x = -3$	Definition of additive inverse
$\frac{1}{4}(4x) = \frac{1}{4}(-3)$	Multiply both sides by $\frac{1}{4}$
$\left(\frac{1}{4} \cdot 4\right)x = \frac{1}{4}(-3)$	Associative property of multiplication
$1x = -\frac{3}{4}$	Multiplicative inverse property
$x = -\frac{3}{4}$	One is a multiplicative identity.

(c) $\frac{2}{3}x + \frac{4}{5} = 0$ Given

$\frac{2}{3}x = -\frac{4}{5}$ Definition of additive inverse

$\frac{3}{2} \cdot \left(\frac{2}{3}x\right) = \frac{3}{2} \cdot \left(\frac{-4}{5}\right)$ Multiply both sides by $\frac{3}{2}$.

$1x = \frac{3 \cdot (-4)}{2 \cdot 5}$ Associative property of multiplication, multiplicative inverse property, definition of multiplication of rational numbers

$x = \frac{-12}{10}$ One is a multiplication identity.

$x = -\frac{6}{5}$ Equivalence of fractions

12. (a) Closure property for subtraction and the existence of a multiplicative inverse, as shown:

$\frac{9}{5}C + 32 = F$

$\left(\frac{9}{5}C + 32\right) - 32 = F - 32$

$\frac{9}{5}C = F - 32$

$\frac{5}{9}\left(\frac{9}{5}C\right) = \frac{5}{9}(F - 32)$

$C = \frac{5}{9}(F - 32)$

13. (a) $-\frac{1}{5}, \frac{2}{5}, \frac{4}{5}$

(number line with points at $-\frac{1}{5}, 0, \frac{2}{5}, \frac{4}{5}, 1$)

(c) $\frac{3}{8}, \frac{1}{2}, \frac{3}{4}$

(number line with points at $0, \frac{3}{8}, \frac{1}{2}, \frac{3}{4}, 1$)

14. Note that in each case the fractions need to be written with positive denominators.

(a) $\frac{-4}{5} < \frac{-3}{4}$: $(-4) \cdot 4 = -16 < -15 = 5 \cdot (-3)$

15. (a) $x + \frac{2}{3} > -\frac{1}{3}$

$x + \frac{2}{3} + \left(-\frac{2}{3}\right) > -\frac{1}{3} + \left(-\frac{2}{3}\right)$

$x > -1$

(c) $\frac{3}{4}x < -\frac{1}{2}$

$\frac{4}{3} \cdot \left(\frac{3}{4}\right) \cdot x < \frac{4}{3} \cdot \left(\frac{-1}{2}\right)$

$x < \frac{-4}{6}$

$x < -\frac{2}{3}$

16. Answers will vary.

(a) One answer is $\frac{1}{2}$, since $\frac{4}{9} < \frac{1}{2} < \frac{6}{11}$.

(c) One answer is the mediant (see Exercise 24 of Problem set 6.1):

$\frac{14+7}{23+12} = \frac{21}{35} = \frac{3}{5}$.

$\frac{14}{23} > \frac{3}{5} > \frac{7}{12}$

18. (a) $\frac{104}{391}$ is approximately $\frac{100}{400} = \frac{1}{4}$.

(c) $\frac{-193}{211}$ is approximately $\frac{-200}{200} = -1$.

19. (a) $3\frac{19}{40} + 5\frac{11}{19}$ is approximately $3\frac{1}{2} + 5\frac{1}{2} = 9$.

20. (a) $\frac{1}{2} + \frac{1}{4} + \frac{3}{4} = \frac{1}{2} + 1 = \frac{3}{2}$

(c) $\frac{3}{4} \cdot \frac{12}{15} = \frac{3}{4} \cdot \frac{4}{5} = \frac{3}{5}$

(e) $2\frac{2}{3} \times 15 = 2 \times 15 + \frac{2}{3} \times 15 = 30 + 10 = 40$

(g) $6\frac{1}{8} - 8\frac{1}{4} = (6-8) + \left(\frac{1}{8} - \frac{1}{4}\right) = -2 - \frac{1}{8}$
$= -2\frac{1}{8}$

21. Estimates will vary.

   (a) 1 (about $\frac{1}{3} + \frac{2}{3}$): $\frac{44,155}{41,866} = 1\frac{2289}{41,866}$.
   Guess was fairly accurate.

   (c) -1 (about $\frac{5}{8} \cdot \frac{-8}{5}$); $\frac{-169}{156} = \frac{-13}{12}$.
   Guess was fairly accurate.

22. (a) [ON/AC] 4 [/] 5 [×] [(] 18 [/] 25 [−] 3 [/] 4 [)]
    [=] [SIMP] [=] [SIMP] [=] , $\frac{-3}{125}$

24.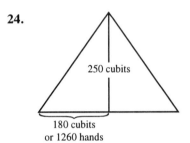

Steepness = $\frac{\text{run in hands}}{\text{rise in cubits}} = \frac{1260}{250} = \frac{126}{25}$
$= 5\frac{1}{25}$

25. (a) $(-4)\cdot 6 + 3\cdot W = 0$
    $3\cdot W = 24$
    $W = 8$

   (c) $(-3)\cdot 2 + x\cdot 10 = 0$
    $10\cdot x = 6$
    $x = \frac{6}{10} = \frac{3}{5}$

   (e) $(-7)\cdot\frac{2}{3} + x\cdot\frac{3}{4} = 0$
    $x\cdot\frac{3}{4} = \frac{14}{3}$
    $x = \frac{14}{3}\cdot\frac{4}{3} = \frac{56}{9}$

26. (a) $s_1 = \frac{1}{1\cdot 2} = \frac{1}{2} = 1 - \frac{1}{2}$
    $s_2 = \frac{1}{1\cdot 2} + \frac{1}{2\cdot 3}$
    $= \frac{1}{2} + \frac{1}{6}$
    $= \left(1 - \frac{1}{2}\right) + \left(\frac{1}{2} - \frac{1}{3}\right)$
    $s_3 = \frac{1}{1\cdot 2} + \frac{1}{2\cdot 3} + \frac{1}{3\cdot 4}$
    $= \frac{1}{2} + \frac{1}{6} + \frac{1}{12}$
    $= \left(1 - \frac{1}{2}\right) + \left(\frac{1}{2} - \frac{1}{3}\right) + \left(\frac{1}{3} - \frac{1}{4}\right)$

27. If $\frac{a}{b} < \frac{c}{d}$, then $\frac{a}{b} + \frac{m}{n} = \frac{c}{d}$ for some rational number $\frac{m}{n} > 0$. If $\frac{c}{d} < \frac{e}{f}$, then $\frac{c}{d} + \frac{r}{s} = \frac{e}{f}$ for some rational number $\frac{r}{s} > 0$. But then,
$\frac{e}{f} = \frac{c}{d} + \frac{r}{s}$
$= \left(\frac{a}{b} + \frac{m}{n}\right) + \frac{r}{s}$
$= \frac{a}{b} + \left(\frac{m}{n} + \frac{r}{s}\right)$.
Hence $\frac{a}{b} < \frac{e}{f}$ since $\frac{m}{n} + \frac{r}{s}$ is a positive rational number.

29. (a) The mediant becomes $\frac{2+2}{4+3} = \frac{4}{7}$, which is not equal to $\frac{3}{5}$.

30. (a) $z = 1$, $y = \frac{4}{5}$, $x = z - y = \frac{1}{5}$,
    $L = x + \frac{1}{2} = \frac{1}{5} + \frac{1}{2} = \frac{7}{10}$

31. (a) Change = $54\frac{1}{2} - 55\frac{1}{4} = \frac{-3}{4}$

   (c) Wed. Close = $46\frac{1}{2} - \left(-\frac{1}{2}\right) = 47$

33. Cut the 8 inch side in half and the 10 inch side in thirds to get six $3\frac{1}{3}$ by 4 inch rectangles.

35. (a) $6\cdot\frac{1}{3} = 2$ square yards

36. One-third off would mean the price is $\frac{2}{3}\cdot 345$ or $230. The maximum price (no savings) would be $345. The price range is $230 to $345.

Classic Conundrum   The Slice Is Right?
(page 425)

Let the 100 loaves be given in the shares s, $s + d$, $s + 2d$, $s + 3d$, and $s + 4d$ to the five men. Then $5s + 10d = 100$, or $s + 2d = 20$. Therefore the shares, again in increasing order, are $20 - 2d$, $20 - d$, 20, $20 + d$, and $20 + 2d$. The other condition of the problem tells us that $40 - 3d = \frac{1}{7}(60 + 3d)$, which can be solved to show that $d = \frac{55}{6} = 9\frac{1}{6}$. Thus the shares are $1\frac{2}{3}$, $10\frac{5}{6}$, 20, $29\frac{1}{6}$, and $38\frac{1}{3}$ loaves.

Chapter 6 Review Exercises (page 427)

1. (a) $\frac{2}{4}$

   (b) $\frac{6}{6}$

   (c) $\frac{0}{4}$

   (d) $\frac{5}{3}$

2.

3. (a) $\frac{27}{81} = \frac{9}{27} = \frac{3}{9} = \frac{1}{3}$

   (b) $\frac{100}{825} = \frac{20}{165} = \frac{4}{33}$

   (c) $\frac{378}{72} = \frac{189}{36} = \frac{63}{12} = \frac{21}{4}$

   (d) $\frac{3^5 \cdot 7^2 \cdot 11^3}{3^2 \cdot 7^3 \cdot 11^2} = \frac{3^3 \cdot 1 \cdot 11^1}{1 \cdot 7^1 \cdot 1} = \frac{297}{7}$

4. $\frac{13}{30}, \frac{13}{27}, \frac{1}{2}, \frac{25}{49}, \frac{26}{49}$

5. (a) $\left\{\ldots, \frac{-15}{-24}, \frac{-10}{-16}, \frac{-5}{-8}, \frac{5}{8}, \frac{10}{16}, \frac{15}{24}, \ldots\right\}$ or $\left[\frac{a}{b} \mid a = 5n,\ b = 8n \text{ for any integer } n \neq 0\right]$

   (b) $\left\{\ldots, \frac{12}{-3}, \frac{8}{-2}, \frac{4}{-1}, \frac{-4}{1}, \frac{-8}{2}, \frac{-12}{3}, \ldots\right\}$ or $\left\{\frac{c}{d} \mid c = -4d,\ \text{where } c \text{ and } d \text{ are integers},\ d \neq 0\right\}$

   (c) $\left\{\ldots, \frac{-9}{-12}, \frac{-6}{-8}, \frac{-3}{-4}, \frac{3}{4}, \frac{6}{8}, \frac{9}{12}, \ldots\right\}$ or $\left\{\frac{e}{f} \mid e = 3k,\ f = 4k \text{ for any integer } k \neq 0\right\}$

6.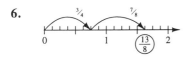

7.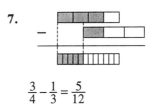

   $\frac{3}{4} - \frac{1}{3} = \frac{5}{12}$

8. (a) $\frac{3}{8} + \frac{1}{4} = \frac{3}{8} + \frac{2}{8} = \frac{5}{8}$

   (b) $\frac{2}{9} + \frac{-5}{12} = \frac{8}{36} + \frac{-15}{36} = \frac{-7}{36}$

   (c) $\frac{4}{5} - \frac{2}{3} = \frac{4 \cdot 3 - 5 \cdot 2}{5 \cdot 3} = \frac{2}{15}$

   (d) $5\frac{1}{4} - 1\frac{5}{6} = \frac{21}{4} - \frac{11}{6} = \frac{63}{12} - \frac{22}{12} = \frac{41}{12}$ or $3\frac{5}{12}$

9. (a)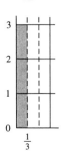

   $3 \times \frac{1}{3} = \frac{3}{3}$ or 1

   (b)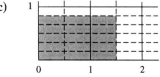

   $\frac{2}{3} \times 4 = \frac{8}{3}$

   (c)

   $\frac{5}{6} \times \frac{3}{2} = \frac{15}{12}$ or $\frac{5}{4}$

10. $7\frac{1}{8} \div 1\frac{1}{4} = \frac{57}{8} \div \frac{5}{4} = \frac{57}{8} \cdot \frac{4}{5} = \frac{57 \cdot 4}{8 \cdot 5}$
    $= \frac{57}{2 \cdot 5} = \frac{57}{10}$

    The distance is $\frac{57}{10}$ miles or $5\frac{7}{10}$ miles.

11. (a) $\frac{-3}{4} + \frac{5}{8} = \frac{-6}{8} + \frac{5}{8} = \frac{-1}{8}$

    (b) $\frac{4}{5} - \frac{-7}{10} = \frac{4}{5} + \frac{7}{10} = \frac{8}{10} + \frac{7}{10} = \frac{15}{10} = \frac{3}{2}$

    (c) $\left(\frac{3}{8} \cdot \frac{-4}{27}\right) \div \frac{1}{9} = \left(\frac{3}{8} \cdot \frac{-4}{27}\right) \cdot \frac{9}{1} = \frac{-4 \cdot 3 \cdot 9}{8 \cdot 27 \cdot 1}$
    $= \frac{-1}{2}$

    (d) $\frac{2}{5} \cdot \left(\frac{3}{4} - \frac{5}{2}\right) = \frac{2}{5} \cdot \left(\frac{3}{4} - \frac{10}{4}\right) = \frac{2}{5} \cdot \frac{-7}{4}$
    $= \frac{-14}{20} = \frac{-7}{10}$

12. (a) $3x + 5 = 11$
    $3x + 5 + (-5) = 11 + (-5)$
    $3x = 6$
    $\frac{1}{3} \cdot (3x) = \frac{1}{3} \cdot 6$
    $x = 2$

    (b) $x + \frac{2}{3} = \frac{1}{2}$
    $x + \frac{2}{3} + \left(-\frac{2}{3}\right) = \frac{1}{2} + \left(-\frac{2}{3}\right)$
    $x = \frac{1}{2} - \frac{2}{3}$
    $x = -\frac{1}{6}$

    (c) $\frac{3}{5}x + \frac{1}{2} = \frac{2}{3}$
    $\frac{3}{5}x + \frac{1}{2} + \left(-\frac{1}{2}\right) = \frac{2}{3} - \frac{1}{2}$
    $\frac{3}{5}x = \frac{1}{6}$
    $\frac{5}{3} \cdot \frac{3}{5}x = \frac{5}{3} \cdot \frac{1}{6}$
    $x = \frac{5}{18}$

    (d) $-\frac{4}{3}x + 1 = \frac{1}{4}$
    $-\frac{4}{3}x + 1 + (-1) = \frac{1}{4} + (-1)$
    $-\frac{4}{3}x = -\frac{3}{4}$
    $\left(-\frac{3}{4}\right) \cdot \left(-\frac{4}{3}\right)x = \left(-\frac{3}{4}\right) \cdot \left(-\frac{3}{4}\right)$
    $x = \frac{9}{16}$

13. Write $\frac{5}{6} = \frac{55}{66}$ and $\frac{10}{11} = \frac{60}{66}$. Then $\frac{56}{66}$ and $\frac{57}{66}$ are between $\frac{5}{6}$ and $\frac{10}{11}$. Other answers may be given.

14. (a) $1\frac{1}{3} + 2\frac{5}{12} + \frac{1}{4} = 1 + 2 + \left(\frac{4}{12} + \frac{5}{12} + \frac{3}{12}\right)$
    $= 1 + 2 + 1 = 4$

    (b) $\frac{6}{7} \cdot \frac{28}{3} \cdot \frac{5}{8} = \frac{6 \cdot 28}{3 \cdot 8 \cdot 7} \cdot 5 = 5$

    (c) $\frac{36}{5} \div \frac{9}{25} = \frac{36}{5} \cdot \frac{25}{9} = \frac{36}{9} \cdot \frac{25}{5} = 4 \cdot 5 = 20$

Chapter 6 Test (page 428)

1. (a)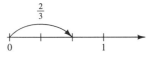

    (b)

    (c)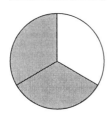

    (d) Answers will vary. One possibility is a set in which $\frac{2}{3}$ of the hearts have arrows through them.

2. Answers will vary. Possibilities include $\frac{-6}{8}, \frac{-9}{12}, \frac{-12}{16}$.

3. $-3, -1\frac{1}{2}, 0, \frac{5}{8}, \frac{2}{3}, 3, \frac{16}{5}$

4. (a) $\frac{1}{3} + \frac{5}{8} - \frac{5}{6} = \frac{8}{24} + \frac{15}{24} - \frac{20}{24} = \frac{3}{24} = \frac{1}{8}$

    (b) $\left(\frac{2}{3} - \frac{5}{4}\right) \div \frac{3}{4} = \left(\frac{8}{12} - \frac{15}{12}\right) \div \frac{3}{4}$
    $= \frac{-7}{12} \div \frac{3}{4} = \frac{-7}{12} \cdot \frac{4}{3} = \frac{-28}{36} = \frac{-7}{9}$

    (c) $\frac{4}{7} \cdot \left(\frac{35}{4} + \frac{-42}{12}\right) = \frac{4}{7} \cdot \left(\frac{35}{4} + \frac{-14}{4}\right)$
    $= \frac{4}{7} \cdot \frac{21}{4} = \frac{21}{7} = 3$

    (d) $\frac{123}{369} \div \frac{1}{3} = \frac{1}{3} \div \frac{1}{3} = 1$

5. (a) If $\frac{a}{b}$ and $\frac{c}{d}$ are rational numbers with $\frac{c}{d} \neq 0$, then $\frac{a}{b} \div \frac{c}{d} = \frac{e}{f}$ if, and only if, $\frac{a}{b} = \frac{c}{d} \cdot \frac{e}{f}$.

    (b) If $\frac{a}{b} \div \frac{c}{d} = \frac{e}{f}$, then $\frac{a}{b} = \frac{c}{d} \cdot \frac{e}{f} = \frac{e}{f} \cdot \frac{c}{d}$.
    So
    $$\frac{a}{b} \cdot \frac{d}{c} = \frac{e}{f} \cdot \frac{c}{d} \cdot \frac{d}{c} = \frac{e}{f}.$$
    Thus, $\frac{a}{b} \div \frac{c}{d} = \frac{a}{b} \cdot \frac{d}{c}$.

6. Answers will vary. One possibility is given in each case.

    (a) Find the area of a rectangular plot of land whose dimensions are $\frac{4}{5}$ mile by $\frac{2}{3}$ mile.

    (b) A drink contains $\frac{3}{10}$ real fruit juice. If a jug of this drink contains $\frac{3}{8}$ of a quart of real juice, how much of the drink is in the jug?

7. (a) $2x + 3 > 0$
    $2x + 3 + (-3) > 0 + (-3)$
    $2x > -3$
    $\frac{1}{2} \cdot 2x > \frac{1}{2} \cdot (-3)$
    $x > \frac{-3}{2}$

    (b) $\frac{3}{4}x + \frac{1}{2} = \frac{1}{3}$
    $\frac{3}{4}x + \frac{1}{2} + \left(-\frac{1}{2}\right) = \frac{1}{3} + \left(-\frac{1}{2}\right)$
    $\frac{3}{4}x = -\frac{1}{6}$
    $\frac{4}{3} \cdot \frac{3}{4}x = \frac{4}{3} \cdot \left(\frac{-1}{6}\right)$
    $x = \frac{-4}{18}$
    $x = \frac{-2}{9}$

    (c) $\frac{5}{4}x > -\frac{1}{3}$
    $\frac{4}{5} \cdot \frac{5}{4}x > \frac{4}{5} \cdot \frac{-1}{3}$
    $x > \frac{-4}{15}$

(d) $\frac{1}{2} < 4x + \frac{5}{6}$

$$4x + \frac{5}{6} > \frac{1}{2}$$

$$4x + \frac{5}{6} + \left(-\frac{5}{6}\right) > \frac{1}{2} + \left(-\frac{5}{6}\right)$$

$$4x > \frac{-1}{3}$$

$$\frac{1}{4} \cdot 4x > \frac{1}{4} \cdot \frac{-1}{3}$$

$$x > \frac{-1}{12}$$

8. $\frac{a+b}{b} = \frac{c+d}{d}$ if, and only if, $(a+b)d = b(c+d)$; that is, if, and only if, $ad + bd = bc + bd$. But this is so if, and only if, $ad = bc$. And this is so if, and only if, $\frac{a}{b} = \frac{c}{d}$.

9. (a) $\frac{1}{8} \cdot \frac{1}{2} = \frac{1}{16}$ square miles

   $\frac{1}{16} \div \frac{1}{640} = \frac{1}{16} \cdot \frac{640}{1} = \frac{640}{16} = 40$ acres

   (b) The area is $80 \cdot \frac{1}{640} = \frac{1}{8}$ square miles, so the width is $\frac{1}{8} \div \frac{1}{2} = \frac{1}{8} \cdot \frac{2}{1} = \frac{1}{4}$ mile.

10. (a) If $\frac{a}{b}$ and $\frac{c}{d}$ are two rational numbers with $\frac{a}{b} < \frac{c}{d}$, then there is a rational number $\frac{e}{f}$ such that $\frac{a}{b} < \frac{e}{f} < \frac{c}{d}$.

    (b) Write $\frac{3}{5} = \frac{18}{30}$ and $\frac{2}{3} = \frac{20}{30}$ to see that $\frac{19}{30}$ is between $\frac{3}{5}$ and $\frac{2}{3}$. Other answers are possible.

11. (a) $2\frac{1}{48} + 3\frac{1}{99} + 6\frac{13}{25}$ is about $2 + 3 + 6\frac{1}{2} = 11\frac{1}{2}$.

    (b) $8 \cdot \left(2\frac{1}{2} + 3\frac{7}{15}\right)$ is about $8 \cdot \left(2\frac{1}{2} + 3\frac{1}{2}\right) = 8 \cdot 6 = 48$.

    (c) $11\frac{9}{10} \div \frac{21}{40}$ is about $12 \div \frac{1}{2} = 12 \cdot 2 = 24$. The closest answer is 23. (Note: the exact value of $11\frac{9}{10} \div \frac{21}{40}$ is $22\frac{2}{3}$, which really is closer to 23 than to 24.)

12. (a) If $\frac{a}{b}$ is a rational number, then its additive inverse is the rational number $\frac{-a}{b}$.

    (b) $\frac{-3}{4}, \frac{7}{4}, \frac{8}{2}$

13. (a) If $\frac{a}{b}$ is a rational number with $a \neq 0$, then its multiplicative inverse is the rational number $\frac{b}{a}$.

    (b) $\frac{2}{3}, -\frac{5}{4}, \frac{-1}{5}$

14. (a) 0, since $\frac{2}{3} + \frac{-4}{6} = 0$.

    (b) 2, since $\frac{5}{6} \cdot \frac{36}{15} = \frac{5}{15} \cdot \frac{36}{6} = \frac{1}{3} \cdot 6 = 2$.

    (c) 1, since $\frac{9}{5} - \frac{1}{5} = \frac{8}{5}$ is the reciprocal of $\frac{5}{8}$.

    (d) $\frac{1}{3}$, since $\frac{2}{3} \cdot \frac{3}{4} \cdot \frac{4}{5} \cdot \frac{5}{6} = \frac{2}{6} = \frac{1}{3}$.

Chapter 7

Problem Set 7.1 (page 449)

1. (a) $273.412 = 200 + 70 + 3 + \frac{4}{10} + \frac{1}{100} + \frac{2}{1000}$;
   $273.412 = 2 \cdot 10^2 + 7 \cdot 10^1 + 3 \cdot 10^0 + 4 \cdot 10^{-1} + 1 \cdot 10^{-2} + 2 \cdot 10^{-3}$

   (b) $0.000723 = \frac{7}{10,000} + \frac{2}{100,000} + \frac{3}{1,000,000}$
   $0.000723 = 7 \cdot 10^{-4} + 2 \cdot 10^{-5} + 3 \cdot 10^{-6}$

   (c) $0.20305 = \frac{2}{10} + \frac{3}{1000} + \frac{5}{100,000}$
   $0.20305 = 2 \cdot 10^{-1} + 3 \cdot 10^{-3} + 5 \cdot 10^{-5}$

2. (a) $0.324 = \frac{324}{1000} = \frac{81}{250}$; $250 = 2 \cdot 5^3$

   Note that, in each case, the denominator is a product of powers of 2 and 5.

3. (a) $\frac{7}{20} = \frac{7 \cdot 5}{20 \cdot 5} = \frac{35}{100} = 0.35$

   (c) $\frac{3}{75} = \frac{1}{25} = \frac{1 \cdot 2^2}{5^2 \cdot 2^2} = \frac{4}{100} = 0.04$

   Note that, in each case, the denominator is a product of powers of 2 and 5.

4. (a) Let $x = 0.321321\ldots$.
   Then $1000x = 321.321321\ldots$
   and $1000x - x = 999x = 321$.
   So $x = \frac{321}{999} = \frac{107}{333}$.

   (d) Let $x = 0.999\ldots$.
   Then $10x = 9.999\ldots$
   and $10x - x = 9x = 9$.
   So $x = \frac{9}{9} = 1$.

   (e) Let $x = 0.24999\ldots$.
   Then $10x = 2.49999\ldots$
   and $10x - x = 9x = 2.25$.
   So $x = \frac{2.25}{9} = \frac{225}{900} = \frac{1}{4}$.

   (g) Let $x = 0.142857142857\ldots$.
   Then $1,000,000x$
   $= 142,857.142857142857\ldots$ and
   $1,000,000x - x = 999,999x = 142,857$.
   So $x = \frac{142,857}{999,999} = \frac{1}{7}$.

5. (a) Answers will vary. For example,
   $\frac{41}{333} = 0.\overline{123}$.

6. (a) $0.007, 0.017, 0.01\overline{7}, 0.027$

   (c) Note that $\frac{9}{25} = 0.36$ and $\frac{10}{25} = 0.40$.
   The numbers in order are
   $0.35, 0.\overline{35}, 0.36 = \frac{9}{25}, 0.40 = \frac{10}{25}$.

8. Assume $3 - \sqrt{2}$ is rational, then $3 - \sqrt{2} = q$ where q is rational. This implies that $\sqrt{2} = 3 - q$. But $3 - q$ is rational since the rational numbers are closed under subtraction. This can't be true since we know $\sqrt{2}$ is irrational. Thus, the assumption that $3 - \sqrt{2}$ is rational must be false. So $3 - \sqrt{2}$ is irrational.

12. (a) Let $x = 0.0\overline{9}$.
    Then $10x = 0.9\overline{9}$ and $10x - x = 9x = 0.9$.
    Therefore, $x = \frac{0.9}{9} = 0.1 = \frac{1}{10}$.

13. Use a calculator or long division for parts (a), (b), and (c).

    (a) $0.\overline{09}$

    (c) $0.\overline{0009}$

14. (a) Let $x = 0.747474\ldots$.
    Then $100x = 74.747474\ldots$
    and $100x - x = 99x = 74$.
    So $x = \frac{74}{99}$.

15. Use the patterns discovered in Exercise 14(d) for parts (a) through (d).

    (a) $\frac{5}{9} = 0.\overline{5}$

    (d) $\frac{17}{33} = \frac{51}{99} = 0.\overline{51}$

16. Answers will vary.

    (a) One example is $\sqrt{2} + (3 - \sqrt{2}) = 3$.
    $\sqrt{2}$ and $(3 - \sqrt{2})$ are both irrational.

    (b) One example is $\sqrt{3} + \sqrt{3} = 2\sqrt{3}$.
    $\sqrt{3}$ and $2\sqrt{3}$ are both irrational.

19. Answers will vary. For example, $\frac{\sqrt{2}}{2\sqrt{2}} = \frac{1}{2}$ and $\frac{1}{2}$ is rational.

25. Answers will vary. If $a = 1$ and $b = 4$, the following results are obtained.

    (a) 1, 4, 5, 9, 14, 23, 37, 60, 97, 157, 254, 411, 665, 1076, 1741, 2817, 4558, 7375, 11933, 19308

27. (a) The medient of $\frac{3}{8}$ and $\frac{17}{24}$ is
    $\frac{3+17}{8+24} = \frac{20}{32}$.

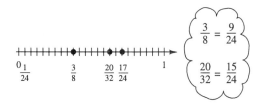

28. (a) $\frac{1}{2} + \frac{2}{3} = \frac{3}{6} + \frac{4}{6} = \frac{7}{6}$

    (c) $\frac{1}{2} \cdot \frac{2}{3} = \frac{1 \cdot 2}{2 \cdot 3} = \frac{1}{3}$

    (e) $\frac{3}{4} \cdot \left(1 + \frac{3}{5}\right) = \frac{3}{4} \cdot \left(\frac{5}{5} + \frac{3}{5}\right) = \frac{3}{4} \cdot \frac{8}{5} = \frac{3 \cdot 8}{4 \cdot 5}$
    $= \frac{3 \cdot 2}{5} = \frac{6}{5}$

Problem Set 7.2 (Page 461)

1. (a) 32.174
    $+371.500$
    $\overline{403.674}$

   (c) 0.057
    $+1.080$
    $\overline{1.137}$

2. (a) 37.1
    $\times4.7$
    $\overline{2597}$
    1484
    $\overline{174.37}$

   (c) $14.664 \div 4.7 = 146.64 \div 47$

    3.12
    $47\overline{)146.64}$
    $\underline{141}$
    56
    $\underline{47}$
    94
    $\underline{94}$
    0

3. (a) Estimate: $4 + 31 = 35$
    Answer: $4.112 + 31.3 = 35.412$

   (c) Estimate: $4 \cdot 31 = 124$
    Answer: $4.112 \cdot 31.3 = 128.7056$

4. (a) $0.833...$
    $6\overline{)5.000}$ $\frac{5}{6} = 0.8\overline{3}$
    $\underline{48}$
    20
    $\underline{18}$
    20

5. (a) 2.77×10^8

6. (a) [ON/AC] 1.27 [EXP] 5 [+/-] [×] 8.235 [EXP] 6 [+/-] [=], 1.05×10^{-10}

   (c) [ON/AC] 1.27 [EXP] 5 [+/-] [+] 98613428 [=], 1.29×10^{-13}

7. (a) Estimate: $(7 \times 10^5) \cdot (2 \times 10^4)$
    $= (7 \cdot 2) \times (10^5 \cdot 10^4) = 14 \times 10^9$
    $= 1.4 \times 10^{10}$
    Answer: $(7.123 \times 10^5) \cdot (2.142 \times 10^4)$
    $\approx 1.53 \times 10^{10}$

8. Answers will vary. Since the given numbers are 0.123 times the digits 1, 2, 3, 4, 5, 6, 7, 8, and 9, respectively, we may obtain a magic square by multiplying each entry in the square in problem 6(a) of Problem Set 1.1 by 0.123. This gives

0.492	1.107	0.246
0.369	0.615	0.861
0.984	0.123	0.738

Aside from rotating and/or flipping the array, this is the only solution.

11. (c) Note that $2.415 - 0.041 = 2.374$,
    $7.723 - 2.415 = 5.308$, and
    $5.308 - 0.041 = 5.267$.

    <pre>
    2.374   0.041   5.267
        2.415   5.308
            7.723
    </pre>

    (e) Yes—(d) can be completed in more than one way.

12. (a) Each number is 0.9 more than its predecessor. The sequence is 3.4, 4.3, 5.2, 6.1, 7.0, 7.9.

    (c) If each number is d more than its predecessor, then $0.0114 + 3d = 0.3204$, so $d = 0.103$. The sequence is 0.0114, 0.1144, 0.2174, 0.3204, 0.4234, 0.5264.

13. (a) Each number is $\frac{2.327}{2.11} = \frac{2327}{2110}$ times its predecessor. With terms rounded to the nearest 0.0001, the sequence is 2.11, 2.327, 2.5663, 2.8302, 3.1213.

15. (a) Note that $1.32 + 3.41 = 4.73$,
    $3.41 + 7.10 = 10.51$, and
    $1.32 + 7.10 = 8.42$.

    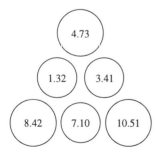

    (c) Recall Example 1.3, page 19. If a, b, and c are the numbers in the small circles, then $2(a + b + c)$

$= (a + b) + (b + c) + (a + c)$
$= 2.341 + 7.133 + 4.012 = 13.486$,
so the sum of a, b, and c is 6.743.
Thus the number opposite 7.133 is
$6.743 - 7.133 = -0.39$ and the other missing numbers are
$6.743 - 2.341 = 4.402$ and
$6.743 - 4.012 = 2.731$.

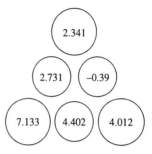

16. (a) Note that $0.92 + 0.41 = 1.33$,
    $0.41 + 1.23 = 1.64$, $1.23 + 0.72 = 1.95$,
    and $0.72 + 0.92 = 1.64$.

    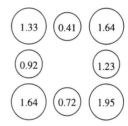

    (c) Answers will vary. A solution may be calculated by placing any number at random in one of the small circles. For example, if 1 is placed in the middle of the bottom row, the remaining entries may be calculated by noting that
    $1.07 - 1 = 0.07$, $-1.41 - 0.07 = -1.48$,
    $2.53 - (-1.48) = 4.01$, and
    $4.01 + 1 = 5.01$. (All possible solutions have 5.01 in the lower right corner.)

    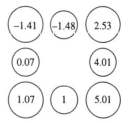

17. $3 \cdot 9.72 = 29.16$, so she spent $29.16.

20. (a) Here it is helpful to draw a picture. Since $24.75 + 2 \cdot 2.25 = 29.25$, the drawing is as shown.

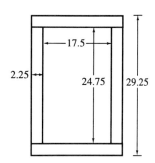

Thus, the area of the frame is
2(2.25 · 17.5) + 2(2.25 · 29.25)
= 210.375 square inches.

21. $\left(\sqrt{3}\right)^2 = 3$ square inches

24. (a)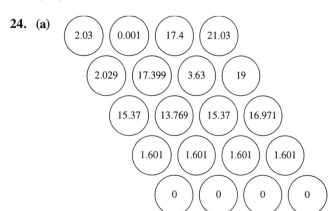

(d) Answers will vary. One possibility is 5, 9, 17, and 31:

```
    5    9   17   31
      4    8   14   26
    4    6   12   22
      2    6   10   18
    4    4    8   16
      0    4    8   12
    4    4    4   12
      0    0    8    8
      0    8    0    8
      8    8    8    8
      0    0    0    0
```

98 *Chapter 7:* Mathematical Reasoning for Elementary Teachers

25. (a) Using the Numeric Surmiser Surpriser disc, the process terminates at the fourth step. Note that when the program prints 1 it means exactly 1; when it prints out 1. it means some number very near 1 like 1.0000002, for example. In this case, the results are as follows:

2.435	3.112	1.024	4.511
1.27803	3.03906	4.40527	1.85257
2.37793	1.44955	2.37793	1.44955
1.64046	1.64046	1.64046	1.64046
1	1	1	1

Note that the number of steps may vary slightly (or even significantly) depending on the accuracy of the calculator or other device being used to perform the calculations.

26. (a) XXIV

27. (a) 29

28. (a) $231 = 11 \cdot 20 + 11$

29. (a) $7 \cdot 7200 + 17 \cdot 360 + 0 \cdot 20 + 5 = 56{,}525$

30. (a)

31. (a) 2303

JUST FOR FUN   Who Shaves Francisco? (page 469)

This is a logical paradox. If Francisco doesn't shave himself, then he must shave himself. However, if he shaves himself then he does not shave himself. This type of self-reference often leads to a paradox and has had to be ruled out in logical discourse.

Problem Set 7.3 (page 472)

1. There are 10 girls, 14 boys, and $10 + 14 = 24$ students.

    (a) boys to girls $= \frac{14}{10} = \frac{7}{5}$

    (c) boys to students $= \frac{14}{24} = \frac{7}{12}$

    (e) students to girls $= \frac{24}{10} = \frac{12}{5}$

2. (a) Yes. $2 \cdot 12 = 3 \cdot 8$

    (c) No. $7 \cdot 31 \neq 28 \cdot 8$

    (e) No. $14 \cdot 60 \neq 49 \cdot 18$

3. (a) $\frac{6}{14} = \frac{r}{21}$
    $6 \cdot 21 = 14 \cdot r$
    $r = \frac{6 \cdot 21}{14} = 9$

    (c) $\frac{51}{t} = \frac{85}{95}$
    $51 \cdot 95 = t \cdot 85$
    $t = \frac{51 \cdot 95}{85} = 57$

4. (a) $\frac{24}{16} = \frac{3}{2}$, so the ratio is 3 to 2.

    (c) $\frac{248}{372} = \frac{2}{3}$, so the ratio is 2 to 3.

5. (a) $3\frac{1}{2} \cdot 5.50 = 19.25$, so she earned $19.25.

    (c) $\frac{3.5}{19.25} = \frac{5}{27.50}$ because
    $3.5 \cdot 27.5 = 96.25$ and $19.25 \cdot 5 = 96.25$.

8. Let x be the height of the flagpole.
    $\frac{x}{6\frac{1}{4}} = \frac{9\frac{2}{3}}{3\frac{1}{6}}$
    $3\frac{1}{6}x = 6\frac{1}{4} \cdot 9\frac{2}{3}$
    $x = \left(6\frac{1}{4} \cdot 9\frac{2}{3}\right) \div 3\frac{1}{6} = \frac{25}{4} \cdot \frac{29}{3} \div \frac{19}{6}$
    $= \frac{25}{4} \cdot \frac{29}{3} \cdot \frac{6}{19}$
    $= \frac{4350}{228} = \frac{725}{38} = 19\frac{3}{38}$
    To the nearest foot, the flagpole is 19 ft high.

10. (a) $\frac{a}{b} = \frac{c}{d}$
    $ad = bc$
    $da = cb$
    $\frac{d}{c} = \frac{b}{a}$

(c) $\dfrac{a}{b} = \dfrac{c}{d}$

$ad = bc$

$ac + ad = ac + bc$

$a(c + d) = (a + b)c$

$\dfrac{a}{a+b} = \dfrac{c}{c+d}$

11. (a) $y = kx^2$

   $27 = k(6^2)$

   $27 = 36k$

   $k = \dfrac{27}{36} = \dfrac{3}{4}$

   When $x = 12$, $y = kx^2 = \dfrac{3}{4}(12^2) = 108$.

12. (a) $y = kx^3$

   $32 = k(12^3)$

   $32 = 1728k$

   $k = \dfrac{32}{1728} = \dfrac{1}{54}$

   When $x = 6$, $y = \dfrac{1}{54}(6^3) = \dfrac{216}{54} = 4$.

15. $\dfrac{100}{11.6} \div \dfrac{100}{11.8} = \dfrac{100}{11.6} \cdot \dfrac{11.8}{100} = \dfrac{1180}{1160} = \dfrac{59}{58}$

   The ratio is 59 to 58.

17. Compare the cost per ounce (or cost per gallon) in each case.

   (a) $\dfrac{0.90}{32} < \dfrac{1.20}{40}$ because

   $0.90 \cdot 40 < 32 \cdot 1.20$, so the better buy is 32 ounces for 90¢.

21. (a) Yes. $2.52 \cdot 18.93 = 7.56 \cdot 6.31$

23. Since the area of a circle is given by the formula $A = \pi r^2$, the price is proportional to the square of the radius of the pizza. Therefore, if P is the price of the large pizza, $\dfrac{P}{7^2} = \dfrac{9.56}{6^2}$.

   Hence, $P = \dfrac{9.56 \cdot 49}{36} = 13.01$.

   The price should be $13.01

26. Let x be the cost of 7 shirts.

   $\dfrac{x}{7} = \dfrac{59.97}{3}$

   $x \cdot 3 = 7 \cdot 59.97$

   $x = \dfrac{7 \cdot 59.97}{3} = 139.93$

   The cost will be $139.93.

29. (a) $4 + (-7) = -(7 - 4) = -3$

   (c) $(-5) + (-7) = -(5 + 7) = -12$

(e) $8 - (-4) = 8 + 4 = 12$

(g) $12 \div (-3) = -(12 \div 3) = -4$

(i) $(-28) \div (-4) = 28 \div 4 = 7$

(k) $57 \div (-19) = -(57 \div 19) = -3$

31. The number of (natural number) divisors is $(5 + 1) \cdot (2 + 1) \cdot (3+1) = 6 \cdot 3 \cdot 4 = 72$.

33. To get the least number divisible by b we multiply b by 2, the least prime.

   $2b = 2^1 \cdot 3^2 \cdot 5 \cdot 11^3$ or 119,790

Problem Set 7.4

1. (a) $100 \cdot \dfrac{3}{16} = 18.75$, so $\dfrac{3}{16} = 18.75\%$.

   (c) $100 \cdot \dfrac{37}{40} = 92.5$, so $\dfrac{37}{40} = 92.5\%$.

   (e) $100 \cdot \dfrac{3.24}{8.91} = 36.\overline{36}$, so $\dfrac{3.24}{8.91} = 36.\overline{36}\%$.

   (g) $100 \cdot \dfrac{1.6}{7} \doteq 22.86$, so $\dfrac{1.6}{7} \doteq 22.86\%$.

2. (a) $100 \cdot 0.19 = 19$, so $0.19 = 19\%$.

   (c) $100 \cdot 2.15 = 215$, so $2.15 = 215\%$.

3. (a) $10\% = \dfrac{10}{100} = \dfrac{1}{10}$

   (c) $62.5\% = \dfrac{62.5}{100} = \dfrac{625}{1000} = \dfrac{5}{8}$

4. (a) $70\% \times 280 = 0.7 \times 280 = 196$

   (c) $38\% \times 751 = 0.38 \times 751 = 285.38$

   (e) $0.02\% \times 27,481 = 0.0002 \times 27,481 = 5.4962$

5. (a) $50\% \times 840 = \dfrac{840}{2} = 420$

   (c) $12.5\% \times 48 = \dfrac{48}{8} = 6$

   (e) $200\% \times 56 = 2 \times 56 = 112$

6. (a) $\dfrac{7}{28} = \dfrac{1}{4} = 25\%$

   (c) $\dfrac{72}{144} = \dfrac{1}{2} = 50\%$

8. Assume that each married man is married to a woman in the same population and vice versa.

If m is the number of men and w is the number of women, then $40\% \times m = 30\% \times w$; that is, $0.4m = 0.3w$ or $m = 0.75w$. The fraction of adults that are married is
$\frac{0.4m + 0.3w}{m + w} = \frac{0.3w + 0.3w}{0.75w + w} = \frac{0.6w}{1.75w} = \frac{0.6}{1.75}$
$\doteq 0.3429$. About 34.29% of the adult population is married.

11. They made $0.6 \times 40 = 24$ field goals in the first half and $0.25 \times 44 = 11$ field goals in the second half. Since $\frac{24 + 11}{40 + 44} = \frac{35}{84} = \frac{5}{12}$
$\doteq 0.42$, their field goal shooting percentage for the game was about 42%

13. (a) Let x be the wholesale cost. Then $2x = x + 100\%x$ is the regular price, and the sale price is $0.8(2x) = 1.6x$. Then the store's profit is $1.6x - x = 0.6x$ and the percent profit is $\frac{0.6x}{x} = 0.60 = 60\%$.

15. $11\% \times 158{,}000 = 0.11 \times \$158{,}000 = \$17{,}380$

17. (a) The markdown is $20\% \times \$78 = \$15.60$, so the sale price is
$\$78 - \$15.60 = \$62.40$.

19. (a) Let $P = 2500$, $r = 5.25$, $t = 1$, and $n = 7$.
Then $P\left(1 + \frac{\frac{r}{t}}{100}\right)^{nt} = 2500\left(1 + \frac{\frac{5.25}{1}}{100}\right)^{7 \cdot 1}$
$= 2500(1.0525)^7 \doteq 3576.80$.
The investment is worth $3576.80.

22. Since $16{,}000 = P \cdot (1.06)^5$,
$P = \frac{16{,}000}{(1.06)^5} \doteq 11{,}956.13$.
You would need to invest $11,956.13.

24. (a) Calculate successive powers of 1.05:
1.05, 1.1025, 1.1576, 1.2155, 1.2763, 1.3401, 1.4071, 1.4775, 1.5513, 1.6289, 1.7103, 1.7959, 1.8856, 1.9799, 2.0789, ...
Since $1.05^{15} > 2$, it takes 15 years.

(c) Calculate successive powers of 1.14:
1.14, 1.2996, 1.4815, 1.6890, 1.9254, 2.1950, ...
It takes 6 years.

25. (a) Since 2010 is 20 years after 1990, the population will be
$2{,}900{,}000(1.04)^{20} \doteq 6{,}400{,}000$.

28. (a) $\frac{51}{69} = \frac{17}{23}$

29. (a) $\frac{1}{12} + \frac{1}{4} = \frac{1}{12} + \frac{3}{12} = \frac{4}{12} = \frac{1}{3}$

(c) $\frac{8}{15} - \frac{17}{45} = \frac{24}{45} - \frac{17}{45} = \frac{7}{45}$

(e) $\frac{11}{84} + \frac{5}{96} = \frac{88}{672} + \frac{35}{672} = \frac{123}{672} = \frac{41}{224}$

30. (a) $\frac{27}{44} \cdot \frac{22}{81} = \frac{27 \cdot 22}{44 \cdot 81} = \frac{1 \cdot 1}{2 \cdot 3} = \frac{1}{6}$

(c) $\frac{33}{35} \div \frac{22}{63} = \frac{33}{35} \cdot \frac{63}{22} = \frac{33 \cdot 63}{35 \cdot 22} = \frac{3 \cdot 9}{5 \cdot 2} = \frac{27}{10}$

(e) $\frac{25}{44} \cdot \frac{68}{75} \cdot \frac{11}{16} = \frac{25 \cdot 68 \cdot 11}{44 \cdot 75 \cdot 16} = \frac{1 \cdot 17 \cdot 1}{1 \cdot 3 \cdot 16} = \frac{17}{48}$

31. (a) $r = \frac{25 \cdot 26}{65} = 10$

CLASSIC CONUNDRUM Some Curious Fractions (page 487)

$$1,\ 2,\ \frac{3}{2},\ \frac{5}{3},\ \frac{8}{5},\ \frac{13}{8}$$

The numbers appear to be quotients of consecutive Fibonacci numbers. Indeed,
$1 = \frac{1}{1} = \frac{F_2}{F_1}$, $2 = \frac{2}{1} = \frac{F_3}{F_2}$,
$\frac{3}{2} = \frac{F_4}{F_3}$, $\frac{5}{3} = \frac{F_5}{F_4}$, $\frac{8}{5} = \frac{F_6}{F_5}$, and $\frac{13}{8} = \frac{F_7}{F_6}$. It appears that the nth term in the sequence is $\frac{F_{n+1}}{F_n}$..
Therefore, the 20th fraction is probably
$\frac{F_{21}}{F_{20}} = \frac{10{,}946}{6765}$.

Chapter 7 Review Exercises (p. 489)

1. (a) $273.425 = 2 \cdot 10^2 + 7 \cdot 10^1 + 3 \cdot 10^0 + 4 \cdot 10^{-1} + 2 \cdot 10^{-2} + 5 \cdot 10^{-3}$

   (b) $0.000354 = 3 \cdot 10^{-4} + 5 \cdot 10^{-5} + 4 \cdot 10^{-6}$

2. (a) $\frac{7}{125} = \frac{56}{1000} = 0.056$

   (b) $\frac{6}{75} = \frac{2}{25} = \frac{8}{100} = 0.08$

   (c) $\frac{11}{80} = \frac{1375}{10,000} = 0.1375$

3. (a) $0.315 = \frac{315}{1000} = \frac{63}{200}$

   (b) $1.206 = \frac{1206}{1000} = \frac{603}{500}$

   (c) $0.2001 = \frac{2001}{10,000}$

4. Since $\frac{4}{12} = 0.\overline{3}$, $\frac{5}{13} \doteq 0.38$, and $\frac{2}{66} \doteq 0.03$, the numbers in order are $0.03, 0.33, 0.\overline{3}, 0.3334, 0.38$, or $\frac{2}{66}, 0.33, \frac{4}{12}, 0.3334, \frac{5}{13}$.

5. (a) Let $x = 10.\overline{363}$.
   Then $1000x = 10,363.\overline{363}$ and $1000x - x = 999x = 10,353$.
   So $x = \frac{10,353}{999} = \frac{3451}{333}$.

   (b) Let $x = 2.1\overline{42}$. Then $100x = 214.2\overline{42}$ and $100x - x = 99x = 212.1$.
   So $x = \frac{212.1}{99} = \frac{2121}{990} = \frac{707}{330}$.

6. Irrational. The decimal expansion of a does not have a repeating sequence of digits and it does not terminate.

7. (a) $0.\overline{2} = \frac{2}{9}$

   (b) $0.\overline{36} = \frac{36}{99} = \frac{4}{11}$

8. (a)
   ```
     21.7340
      3.2145
    +71.2400
     96.1885
   ```

   (b)
   ```
     23.471
     -2.890
     20.581
   ```

   (c)
   ```
       35.4
     ×2.37
      2478
      1062
       708
     83.898
   ```

   (d) $24.15 \div 3.45 = 2415 \div 345 = 7$
   ```
          7
     345)2415
         2415
            0
   ```

9. (a) $31.47 + 3.471 + 0.0027 = 34.9437$

   (b) $31.47 - 3.471 = 27.999$

   (c) $31.47 \times 3.471 = 109.23237$

   (d) $138.87 \div 23.145 = 6$

10. Estimates may vary:

    (a) Estimate: $50 + 10 = 60$
    Answer: $47.25 + 13.134 = 60.384$

    (b) Estimate: $50 - 10 = 40$
    Answer: $52.914 - 13.101 = 39.813$

    (c) Estimate: $50 \times 10 = 500$
    Answer: $47.25 \times 13.134 = 620.5815$

    (d) Estimate: $45 \div 15 = 3$
    Answer: $47.25 \div 13.134 \doteq 3.5975$

11. (a) $24,732,654 \doteq 2.473 \times 10^7$

    (b) $0.000012473 \doteq 1.247 \times 10^{-5}$

12. (a) $(2.74 \times 10^5) \cdot (3.11 \times 10^4)$
    $= 8.5214 \times 10^9 \doteq 8.52 \times 10^9$

    (b) $(2.74 \times 10^5) \div (3.11 \times 10^{-4}) \doteq 8.81 \times 10^8$

13. Suppose $3-\sqrt{2} = r$ where r is rational. Then $3-r=\sqrt{2}$. But this implies that $\sqrt{2}$ is rational since the rationals are closed under subtraction. This is a contradiction since $\sqrt{2}$ is irrational. Therefore, by contradiction, $3-\sqrt{2}$ is irrational.

14. Answers may vary. $3-\sqrt{2}$ and $\sqrt{2}$ are irrational, but $(3-\sqrt{2})+\sqrt{2} = 3$, which is rational.

15. The decimal expansion of an irrational number is not repeating and is nonterminating.

16. (a) $8.25 \times 112.5 = 928.125$ ft^2

    (b) $928.125 \div 110 = 8.4375$ qts

17. Answers will vary. For example, $\frac{4123}{9999}$ is such a fraction.

18. (a) $\frac{5}{18} \doteq 0.2777778$, so $\frac{5}{18} = 0.2\overline{7}$; the period starts in the second decimal place.
    $\frac{41}{333} \doteq 0.1231231$, so $\frac{41}{333} = 0.\overline{123}$; the period starts right after the decimal point, in the first decimal place.
    $\frac{11}{36} \doteq 0.3055556$, so $\frac{11}{36} = 0.30\overline{5}$; the period starts in the third decimal place.
    $\frac{7}{45} \doteq 0.1555556$, so $\frac{7}{45} = 0.1\overline{5}$; the period starts in the second decimal place.
    $\frac{13}{80} = 0.1625$; this decimal terminates but can be written as $0.1625\overline{0}$ or $0.1624\overline{9}$, so one might say that the period starts in the fifth decimal place.

    (b) Look for a pattern based on the prime factorization of the denominator:

Fraction	$\frac{5}{18}$	$\frac{41}{333}$	$\frac{11}{36}$	$\frac{7}{45}$	$\frac{13}{80}$
Factorization	$2^1 \cdot 3^2$	$3^2 \cdot 37$	$2^2 \cdot 3^2$	$3^2 \cdot 5^1$	$2^4 \cdot 5$
Period starts at	2	1	3	2	5

    The period begins in decimal place $r + 1$, where r is the highest power of 2 and/or 5 appearing in the prime factor representation of the denominator of the fraction in simplest form.

19. She had 11 successes and $20 - 11 = 9$ failures, so the ratio is 11 to 9.

20. (a) Yes. $775 \cdot 25 = 125 \cdot 155$

    (b) No. $31 \cdot 32 \neq 64 \cdot 15$

    (c) Yes. $9 \cdot 32 = 24 \cdot 12$

21. Let x be the cost of 5 pounds of candy.
    $\frac{x}{5} = \frac{3.15}{2}$
    $x = \frac{5 \cdot 3.15}{2} = 7.875$
    It would cost $7.88.

22. Let x be the number of gallons used to travel 300 miles.
    $\frac{x}{300} = \frac{7.5}{173}$

$x = \dfrac{300 \cdot 7.5}{173} \doteq 13.006$

He will need about 13 gallons.

23. $y = kx$
    $7 = k \cdot 3$
    $k = \dfrac{7}{3}$
    If $x = 5$, $y = kx = \dfrac{7}{3} \cdot 5 = \dfrac{35}{3}$.

24. Let x be the height of the flagpole in feet, and note that the yardstick has length 3 feet and its shadow has length $\dfrac{5}{6}$ foot.
    $\dfrac{x}{3} = \dfrac{12}{\frac{5}{6}} = \dfrac{12 \cdot 6}{5}$
    $x = 3 \cdot \dfrac{12 \cdot 6}{5} = 43.2$
    The flagpole is 43.2 feet tall.

25. (a) $\dfrac{5}{8} = 0.625 = 62.5\%$

    (b) $2.115 = 211.5\%$

    (c) $0.015 = 1.5\%$

26. (a) $28\% = \dfrac{28}{100} = 0.28$

    (b) $1.05\% = \dfrac{1.05}{100} = 0.0105$

    (c) $33\tfrac{1}{3}\% = \dfrac{33\tfrac{1}{3}}{100} = \dfrac{1}{3} = 0.\overline{3}$

27. $7.2\% \times \$49 = 0.072 \times \$49 \doteq \$3.53$

28. $\dfrac{6.75}{84.37} \doteq 0.08 = 8\%$

29. $\dfrac{11}{20} = 0.55 = 55\%$

30. $\$3000\left(1 + \dfrac{\frac{8}{4}}{100}\right)^{2 \cdot 4} = \$3000(1.02)^8$
    $\doteq \$3514.98$

Chapter 7 Test (page 490)

1. (a) $\dfrac{84}{175} = \dfrac{12}{25} = \dfrac{48}{100} = 0.48$

   (b) $\dfrac{24}{99} = 0.\overline{24}$

   (c) $\dfrac{7}{11} = \dfrac{63}{99} = 0.\overline{63}$

2. (a) Let $x = 0.\overline{45}$.
       Then $100x = 45.\overline{45}$ and $100x - x = 99x = 45$. So $x = \dfrac{45}{99} = \dfrac{5}{11}$.

   (b) Let $x = 31.\overline{5}$. Then $10x = 315.\overline{5}$ and $10x - x = 9x = 284$. So $x = \dfrac{284}{9}$.

   (c) Let $x = 0.34\overline{9}$. Then $10x = 3.49\overline{9}$ and $10x - x = 9x = 3.15$.
       So $x = \dfrac{3.15}{9} = \dfrac{315}{900} = \dfrac{7}{20}$.

3. Since 21.432 has 3 digits to the right of the decimal point and 3.41 has 2 digits to the right of the decimal point, the number of digits to the right of the decimal point in the product should be $3 + 2 = 5$.

4. $(2.34 \times 10^6) \cdot (3.12 \times 10^{-19}) = 7.3008 \times 10^{-13}$
   $\doteq 7.30 \times 10^{-13}$

5. Answers will vary. A suitable choice is $0.\overline{125} = \dfrac{125}{999}$.

6. (a) Since there are 17 wins and $32 - 17 = 15$ losses, the ratio is 17 to 15.

   (b) $\dfrac{17}{32} = 0.53125 = 53.125\%$

7. $\dfrac{1425}{9500} = 0.15 = 15\%$

8. Your investment is multiplied by 1.021 every six months. By keying in
   2000 ⊠ 1.021 ⊟ ⊟ ⊟ ..., it takes 34 presses of the ⊟ key to obtain a number over 4000 (namely, 4054.20). Therefore, it will take 34 half-years, or 17 years.

9. Your investment is multiplied by 1.0575 every year. By keying in
   5000 ⊠ 1.0575 ⊟ ⊟ ⊟ ⊟, you will find that it takes five presses of the ⊟ key to obtain approximately $6612.60. It takes five years, so you invested the money 5 years ago.

10. $\$1000\left(1 + \dfrac{\frac{9}{12}}{100}\right)^{2 \cdot 12} = \$1000(1.0075)^{24}$
    $= \$1196.41$

Chapter 8

JUST FOR FUN A Matter of Speed (p. 496)

To drive 300 miles at an average speed of 60 miles per hour requires $300 \div 60 = 5$ hours. Since it takes 5 hours to drive the first 250 miles at 50 miles per hour, it is impossible to average 60 miles per hour for the entire trip.

Problem Set 8.1 (page 505)

3. Count the number of scores in each of the intervals to obtain the frequency and thus the height of the bar above the interval. For example, the first interval (20–29) has no scores so the height is 0. The second interval (30–39) has 1 score so the height above the interval is 1.

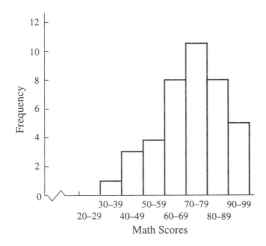

7. (a) Draw the vertical scale from 10 to 73 by 10 and label the axis "million." Draw the horizontal scale with 10 equal intervals and label every other year. Use the data to draw the two line graphs on the axes and label each one.

(b) To find the average number of tons of milk produced per cow, divide the milk production by the number of cows.

1980: $\dfrac{64.2}{10.8} \doteq 5.9$

1989: $\dfrac{72.6}{10.1} \doteq 7.2$

(c) It is no doubt due to better production practices on the part of dairy farmers—primarily better feeding practices, better health care for cows, and perhaps to the use of hormones.

9. Find the central angle to the nearest degree, for each expense.

Taxes: $\dfrac{210{,}000}{64{,}000} \cdot 360° \doteq 118°$

Rent: $\dfrac{10{,}800}{64{,}000} \cdot 360° \doteq 61°$

Food: $\dfrac{5{,}000}{64{,}000} \cdot 360° \doteq 28°$

Clothes: $\dfrac{2{,}000}{64{,}000} \cdot 360° \doteq 11°$

Car payments: $\dfrac{4{,}800}{64{,}000} \cdot 360° \doteq 27°$

Insurance: $\dfrac{5{,}200}{64{,}000} \cdot 360° \doteq 29°$

Charity: $\dfrac{7{,}000}{64{,}000} \cdot 360° \doteq 39°$

Savings: $\dfrac{6{,}000}{64{,}000} \cdot 360° \doteq 34°$

Misc: $\dfrac{2{,}200}{64{,}000} \cdot 360° \doteq 12°$

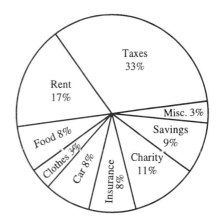

11. (a) Looking at the height of the bar for the C grade, 12 students received a C.

12. (a) In 1974, roughly $5 billion out of the total of $7 billion in export earnings came from the sale of oil and natural gas, or $\dfrac{5}{7} \cdot 100\% \doteq 71\%$.

16. Consider the vertical line through 2040. Then draw or imagine horizontal lines through the points where this line intersects the solid red and dotted red graphs. Where the lines intersect the vertical axis gives the desired approximations as follows:

(a) approximately 139 million

17. (a) The red line above the year 1983 indicates about 310 arrests for violent crimes per 100,000 juveniles.

21. (a) Histogram (A) emphasizes the changes in the daily Dow Jones average by its choice of vertical scale. The changes appear to be large in histogram (A) thus exaggerating the report of stock activity on the evening news. Histogram (B) makes it clear that the changes are minimal.

(b) $\left(\frac{36}{3086}\right) \cdot (100\%) \doteq 1.2\%$ which an investor probably would not worry about.

24. (b) First can:
$V = \pi (4 \text{ cm})^2 (8 \text{ cm}) \doteq 402 \text{ cm}^3$
Second can:
$V = \pi (3 \text{ cm})^2 (14 \text{ cm}) \doteq 396 \text{ cm}^3$

25. (b) The graph is difficult to interpret since the points plotted indicate the percentage increase month by month over the preceding years. From January through June the increase was roughly 3% per month, from July through September it was about 6% per month, and from October through December it was between 7% and 10% per month.

26. (a) Expenditures per elementary or secondary school student
1960: approximately
$\frac{\$80,000,000,000}{40,000,000} \doteq \$2000$
1990: approximately
$\frac{\$230,000,000,000}{46,000,000} \doteq \$5000$

(d) The graph of the data in current dollars exaggerates the increase in spending since it also shows growth due to inflation. That is, if a dollar is worth less in 1990 than in 1960, you have to spend more in 1990 just to stay at the same relative level of spending. The graph of the data in constant 1990 dollars is adjusted for inflation and so gives a more realistic picture of the increase in spending.

27. (a) Since 22.5% of the people are not high school graduates, the percentage of people who graduated from high school is 100% − 22.5% = 77.5%. (Note that we assume that those who attended college or earned a higher degree are high school graduates.) The percentage desired in the question is the fraction (expressed as a percentage) of the 77.5% of those who graduated from high school who go to college but fail to get a bachelor's degree. This would include those who attend college but do not obtain a degree, those who obtain vocational technical certificates, and those who obtain associate degrees. Thus, the desired percentage is
$\frac{17.6 + 2.1 + 4.2}{77.5} \doteq 0.31 = 31\%$.

28. (a) The largest percentage overall (total column) is 33% which corresponds to the problem of students drinking alcohol.

36. (a) $\frac{1}{8} = \frac{1}{2^3} = \frac{1 \cdot 5^3}{2^3 \cdot 5^3} = \frac{5^3}{10^3} = \frac{125}{1000} = 0.125$

(c) $\frac{17}{250} = \frac{17}{2 \cdot 5^3} = \frac{17 \cdot 2^2}{2^3 \cdot 5^3} = \frac{17 \cdot 4}{10^3} = \frac{68}{1000}$
$= 0.068$

37. (a) $\frac{1}{9} = 0.1111... = 0.\overline{1}$

(c) $\frac{3}{14} = 0.2142857142857... = 0.2\overline{142857}$

38. (a) $0.375 = 0.375 \cdot \frac{1000}{1000} = \frac{375}{1000} = \frac{3}{8}$

(c) $0.444 = 0.444 \cdot \frac{1000}{1000} = \frac{444}{1000} = \frac{111}{250}$

39. (a) If $x = 3.7\overline{4}$, then
$100x = 374.\overline{4}$ and $10x = 37.\overline{4}$.
$100x - 10x = 90x = 337$ so $x = \frac{337}{90}$.
$3.7\overline{4} = \frac{337}{90}$

(c) If $x = 0.0\overline{2}$, then
$100x = 2.\overline{2}$ and $10x = 0.\overline{2}$.
$100x - 10x = 90x = 2$ so $x = \frac{2}{90} = \frac{1}{45}$.
$0.0\overline{2} = \frac{1}{45}$

(e) If $x = 4.7\overline{314}$, then
$10,000x = 47,314.\overline{314}$ and
$10x = 47.\overline{314}$.
$10,000x - 10x = 9990x = 47,267$ so
$x = \frac{47,267}{9990}$.
$4.7\overline{314} = \frac{47,267}{9990}$

JUST FOR FUN **How Many are Single?**
(page 516)

Let m be the number of men and let w be the number of women in the population. Assume that every married man is married to a woman in the same population and vice versa. Then $\frac{3}{5}m = \frac{4}{7}w$, so $m = \frac{20}{21}w$. Since $\frac{2}{5}$ of the men and $\frac{3}{7}$ of the adult population are single, the single adult population is
$\frac{2}{5} \cdot m + \frac{3}{7} \cdot w = \frac{2}{5} \cdot \left(\frac{20}{21}w\right) + \frac{3}{7}w$
$= \frac{8}{21}w + \frac{9}{21}w = \frac{17}{21}w$, and the total adult population is $m + w = \frac{20}{21}w + w = \frac{41}{21}w$. The fraction that are single is $\frac{17}{21}w \div \frac{41}{21}w = \frac{17}{21} \div \frac{41}{21}$
$= \frac{17}{21} \cdot \frac{21}{41} = \frac{17}{41}$.

Problem Set 8.2 (page 527)

1. mean: $\overline{x} = \dfrac{\sum_{i=1}^{n} x_i}{n} = \dfrac{\sum_{i=1}^{16} x_i}{16} = \dfrac{18 + 27 + \cdots + 27 + 30}{16} = \dfrac{329}{16} = 20.5625 \doteq 20.6$

   median: arrange the values in order from smallest to largest
   14 15 17 17 18 18 18 19 19 21 22 23 24 27 27 30
   The median is the average of the two middle values. $\hat{x} = \dfrac{19 + 19}{2} = 19$

   mode: The mode is the value that occurs most often. In this case, the mode is 18.

3. (a) Listing the values from Problem 2 in order: 64, 67, 67, 69, 69, 69, 70, 70 70, 71, 71, 73, 74, 77, 77, 78, 79, 79, 79, 79, 80, 80, 81, 81, 81, 85, 86
   Q_L is the median for the first 13 values: $Q_L = 70$.
   $\hat{x}$ is the median of the entire set: $\hat{x} = 77$.
   Q_U is the median for the last 13 values: $Q_U = 80.$.

   (b) The 5-number summary is: the smallest value – Q_L – $\hat{x}$ – Q_U – the largest value; i.e.,
   64 – 70 – 77 – 80 – 86.

5. (a) $\overline{x} = \dfrac{\sum_{i=1}^{16} x_i}{16} = \dfrac{329}{16} = 20.5625 \doteq 20.56$

   (b) $s = \sqrt{\dfrac{\sum_{i=1}^{16}(\overline{x} - x_i)^2}{16}} = \sqrt{\dfrac{(20.5625 - 18)^2 + (20.5625 - 27)^2 + \cdots + (20.5625 - 30)^2}{16}}$
   $\doteq \sqrt{\dfrac{6.5664 + 41.4414 + \cdots + 89.0664}{16}} \doteq \sqrt{\dfrac{315.9375}{16}} \doteq 4.44$

(c) To be within one standard deviation of the mean, a data value must be in the interval $(20.56 - 4.44, 20.56 + 4.44) = (16.12, 25)$.
There are 11 out of 16 values between 16.12 and 25. $\frac{11}{16} \doteq 0.69 = 69\%$

(d) To be within two standard deviations of the mean, a data value must be in the interval $(20.56 - (2) \cdot (4.44), 20.56 + (2) \cdot (4.44)) = (11.68, 29.44)$
There are 15 out of 16 values between 11.68 and 29.44. $\frac{15}{16} \doteq 0.94 = 94\%$

(e) To be within three standard deviations of the mean, a data value must be in the interval $(20.56 - (3) \cdot (4.44), 20.56 + (3) \cdot (4.44)) = (7.24, 33.88)$.
All 16 values are between 7.24 and 33.88: $\frac{16}{16} = 1 = 100\%$

7. (a) $\sum_{i=3}^{8} i^2 = 3^2 + 4^2 + 5^2 + 6^2 + 7^2 + 8^2$ or $9 + 16 + 25 + 36 + 49 + 64$

8. (a) $\sum_{i=1}^{1} \frac{1}{i(i+1)} = \frac{1}{1 \cdot (1+1)} = \frac{1}{2}$

(b) $\sum_{i=1}^{2} \frac{1}{i(i+1)} = \frac{1}{1 \cdot (1+1)} + \frac{1}{2 \cdot (2+1)} = \frac{1}{2} + \frac{1}{6} = \frac{3+1}{6} = \frac{4}{6} = \frac{2}{3}$

9. (a) $\left(\sum_{i=1}^{5} i\right)^2 = (1+2+3+4+5)^2 = 15^2 = 225$

10. (a) Notice the sequence is an arithmetic progression where each term is 2 larger than its predecessor.
$3 = 3, 5 = 3 + 1 \cdot 2, 7 = 3 + 2 \cdot 2, 9 = 3 + 3 \cdot 2, 11 = 3 + 4 \cdot 2 \Rightarrow a_i = 3 + (i-1)2 = 2i+1$ where i starts at 1 and increases to 5. The sequence may be represented as $\sum_{i=1}^{5} (2i+1)$. Other answers are possible, such as $\sum_{j=0}^{4} (2j+3)$.

(c) Notice that the sequence is a geometric sequence where each term is 3 times larger than its predecessor.
$2 = 2, 6 = 2 \cdot 3, 18 = 2 \cdot 3^2, 54 = 2 \cdot 3^3, 162 = 2 \cdot 3^4, 486 = 2 \cdot 3^5, 1458 = 2 \cdot 3^6$. Therefore, $a_i = 2 \cdot 3^i$ where i starts at 0 and increases to 6. $\sum_{i=0}^{6} 2 \cdot 3^i$

11. June 1, 1993: $n = 33, \bar{x} = 47 \Rightarrow \sum x = n \cdot \bar{x} = 1551$
June 1, 1994: $\sum x = 1551 + 33 = 1584$
staff change: $1584 - (65 + 58 + 62) + (24 + 31 + 26 + 28) = 1508$
average age: $\bar{x} = \frac{1508}{34} \doteq 44.4$ years

12. (a) mean: $\bar{x} = \dfrac{\sum\limits_{i=1}^{n} x_i}{n} = \dfrac{28+34+41+19+17+23}{6} = \dfrac{162}{6} = 27$

standard deviation:

$s = \sqrt{\dfrac{\sum\limits_{i=1}^{n}(\bar{x}-x_i)^2}{n}} = \sqrt{\dfrac{(27-28)^2+(27-34)^2+(27-41)^2+(27-19)^2+(27-17)^2+(27-23)^2}{6}}$

$= \sqrt{\dfrac{1+49+196+64+100+16}{6}} = \sqrt{\dfrac{426}{6}} \doteq 8.4$

(b) mean: $\bar{x} = \dfrac{33+39+46+24+22+28}{6} = \dfrac{192}{6} = 32$

standard deviation:

$s = \sqrt{\dfrac{(32-33)^2+(32-39)^2+(32-46)^2+(32-24)^2+(32-22)^2+(32-28)^2}{6}} = \sqrt{\dfrac{426}{6}} \doteq 8.4$

15. (a) $\bar{x} = \dfrac{1\cdot12+1\cdot18+5\cdot22+1\cdot24+3\cdot26+2\cdot28+4\cdot30+3\cdot32+1\cdot36+2\cdot38+2\cdot42+1\cdot60+3\cdot100}{29} \cdot 1000$

$= \dfrac{1070}{29} \cdot 1000 \doteq 36{,}900$

To find the median, find the 15th ordered value: $\hat{x} = 30{,}000$. The mode is 22,000.

16. Answers will vary.

(a) Consider the condition: mean = median

Suppose $\bar{x} = \hat{x} = 10$. Now choose 2 values below $\bar{x}$, say 5 and 7. But we must also choose 2 values above $\bar{x}$ so that $\dfrac{\sum x_i}{n} = \bar{x}$, say 13 and 15. The condition mean = median has been met with the data: 5, 7, 10, 13, 15.

Consider the condition: mean = median < mode.

Using the data above, there must be 2 equal values larger than 10 to meet this condition. If 13 and 15 are replaced with their average of 14, then mean = median < mode for the data: 5, 7, 10, 14, 14.

For these values, $\bar{x} = 10$, $\hat{x} = 10$, and the mode is 14.

17. (a) If $s = \sqrt{\dfrac{\sum\limits_{i=1}^{n}(\bar{x}-x_i)^2}{n}} = \sqrt{\dfrac{(\bar{x}-x_1)^2+(\bar{x}-x_2)^2+\cdots+(\bar{x}-x_n)^2}{n}} = 0$, then $\sum\limits_{i=1}^{n}(\bar{x}-x_i)^2 = 0$ and $(\bar{x}-x_i)^2 = 0$ for every x_i. That means that $\bar{x} = x_i$ for every x_i. Therefore, all the data values are equal.

19. (a) Given $n = 10$ and $\bar{x} = 3$, then $\sum x = n \cdot \bar{x} = 10 \cdot 3 = 30$. Since $\sum x = 30$ for 10 values of ones, twos, and threes, the 10 values must all be threes.

20. (a) $\bar{x}_A = \dfrac{27+38+25+29+41}{5} = 32$

(c) $\bar{x}_C = \dfrac{27+38+25+29+41+32+32}{7} = 32$

23. The total of data values in A is $30 \cdot 45 = 1350$. The total of data values in B is $40 \cdot 65 = 2600$. For the combined data, $\bar{x} = \dfrac{1350+2600}{30+40} \doteq 56.4$.

27. (a) Estimate value from the bar graph.

Month	O	N	D	J	F	M	A	M	J	J	A	S
Deficit (in billions)	7.2	7.8	7.0	7.7	7.9	10.4	10.0	8.4	12.0	10.4	10.1	10.0	9

$\bar{x} = 9.15$ billion

29. (a) $\bar{x} \doteq 76.17$

   (b) $\hat{x} = 78$

   (c) the mode = 79 or 78

   (d) $s \doteq 11.53$

   (e) $Q_L = 68$

   (f) $Q_U = 86$

31. There are many possible answers. One pair would be $\frac{11}{16}$ and $\frac{21}{32}$.

33. Assume that $\sqrt{3} + r = s$ where s is rational. Then $\sqrt{3} = s - r$, and $s - r$ is rational since the rational numbers are closed under subtraction. But this says $\sqrt{3}$ is rational and is a contradiction. Therefore, $\sqrt{3} + r$ is irrational.

JUST FOR FUN   **A Matter of Sums (page 534)**

We guess and check.

$$1 + 23 + 4 + 5 + 67 + 8 + 9 = 117$$
$$12 + 3 + 4 + 5 + 67 + 8 + 9 = 108$$

This reduces the sum by 9. If we could reduce it by 9 more, we would be done. Perhaps we can use the same idea; that is, replace 5 + 67 by 56 + 7. This gives

$$12 + 3 + 4 + 56 + 7 + 8 + 9 = 99$$

as desired.

Notice that any number we obtain in this way must differ from 117 by a multiple of 9. An interesting study would be to determine *all* numbers that can be obtained in the way described.

Problem Set 8.3 (page 540)

1. (a) All freshmen in U.S. colleges and universities in 1995

   (c) All people in the U.S.

2. Yes. Many poorer people cannot afford telephones, many people are irritated by telephone surveys and sales pitches, and so on. These factors could certainly bias a sample.

5. (a) This is surely a poor sampling procedure. The sample is clearly not random. The selection of the colleges or universities could easily reflect biases of the investigators. The choices of the faculty to be included in the study almost surely also reflects the bias of the administrators of the chosen schools.

   (b) Presumably the population is all college and university faculty. But the opinions of faculty at large research universities are surely vastly different from those of their colleagues at small liberal arts colleges. Indeed, there are almost surely four distinct populations here.

9. (a) Suppose the container contains n beans. Then the number of marked beans is $\frac{25}{n}$. This fraction would be approximated by the fraction $\frac{a}{b}$ of marked beans in the handful. Thus,

   $$\frac{25}{n} \doteq \frac{a}{b} \text{ and } n \doteq \frac{25b}{a}$$

   is approximately the number of beans in the container.

10.
   |←s→|←s→|←s→|̄x̄|←s→|←s→|←s→|
   16.4 19.1 21.8 24.5 27.2 29.9 32.6

   (a) The limit 2.7 units less than 24.5 is 21.8. The limit 2.7 units more than 24.5 is 27.2.

11. Yes, since all sides of the die are equally likely to come up, all sequences of 0s and 1s are equally likely to appear.

15. (b) Geometric mean $= \sqrt[5]{1 \cdot 2 \cdot 4 \cdot 8 \cdot 16}$
$= (1024)^{1/5} = 4$

16. (a) Let d be the distance traveled in time t at the constant rate of speed r. Then $rt = d$ and $d \div r = t$. Thus, Katja's time going upstream is $4 \div 2 = 2$ hours and her time going downstream is $4 \div 4 = 1$ hour. Thus, for the total trip, the average speed is $8 \div (2 + 1) = \frac{8}{3}$
$= 2.\overline{6}$ miles per hour.

(b) The harmonic mean of 2 and 4 is the reciprocal of the average of their reciprocals; i.e.,
$\frac{1}{\frac{\frac{1}{2}+\frac{1}{4}}{2}} = \frac{2}{\frac{1}{2}+\frac{1}{4}} = \frac{8}{2+1} = \frac{8}{3}.$

23. (a) $\bar{x} = \frac{44.3}{10} = 4.43$

(b) $s = \sqrt{\frac{(4.43 - 3.4)^2 + (4.43 - 4.4)^2 + \cdots + (4.43 - 3.6)^2}{10}} = \sqrt{\frac{4.561}{10}} \doteq 0.68$

27. (a) $\bar{x} = \frac{115}{5} = 23$

(b) $\bar{x} = \frac{138}{6} = 23$

CLASSIC CONUNDRUM   **A Magic Magic Magic Square (page 544)**

There are many patterns that all add to 34. Some not already shown include the following.

Chapter 8 Review Exercises (page 545)

1. (a)
```
                    x
                x   x
            x   x x       x
        x x x   x x x x   x
        x x x x x x x x x x   x   x
        ─┬───────┬───────┬───────→
         0       10      20
```

(b) About 13

2. 0 | 7 7 8 8 8 9 9
    1 | 0 1 1 1 1 2 2 2 3 3 3 3 4 4 5 6 7 7 7 9
    2 | 1

3. Use a bar for each number of hours.

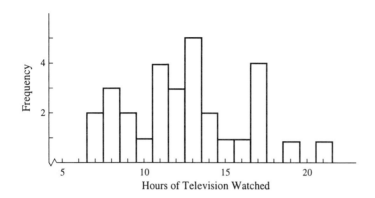

4. Put Mrs. Karnes class on the left of the stems and Ms. Stevens class on the right of the stems.

Mrs. Karnes		Ms. Stevens
9 9 8 8 8 7 7	0	6 6 6 8 8 8 8 8 9 9 9 9 9
9 7 7 7 7 6 5 4 4 3 3 3 3 3 2 2 2 1 1 1 1 0	1	1 1 1 1 1 1 1 1 2 3
	2	

5. (a) The vertical axis should be the Retail Price Index scale, and the horizontal axis should have years marked on the scale.

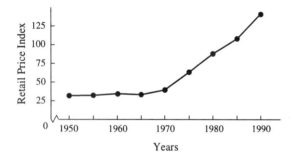

(b) Draw a line straight up from the year 1972 on the horizontal axis. Now draw a horizontal line from the point on the line graph above 1972 to the vertical axis. The retail price index for farm products in 1972 was about 50.

(c) Assuming the retail prices index continues the same upward trend as that from 1970 to 1990, the retail price index for farm products in 2000 will be about 180. If the trend from 1990 to 2000 differs from the trend from 1970 to 1990, the estimate of 180 may not be very accurate.

6. Many examples exist. One example is 1, 2, 3, 4, 90 with a mean of 20.

7. Find the central angle for each section of the budget.
Administration: 12% of 360° = 43.2°
New construction: 36% of 360° = 129.6°
Repairs: 48% of 360° = 172.8°
Miscellaneous: 4% of 360° = 14.4°

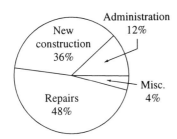

8. (a) The volume of the larger box is about double that of the smaller box. But the length of a side of the larger box is less than double the length of a side of the smaller box, suggesting that the change was less than doubling.

   (b) The smaller cube has side length of 18 mm and volume of 5832 cubic mm. The larger cube has side length of 22.5 mm and volume of about 11,500 cubic mm. The volume of the larger cube is about twice that of the smaller cube, which defends the pictograph.

9. How is the medical doctor defined? Does this include all specialists? osteopaths? naturopaths? chiropractors? acupuncturists? How was the sampling done to determine the stated average?

10. $\bar{x} = \frac{378}{30} = 12.6$

    The median is the average of the 15th and 16th ordered values.
    $\hat{x} = \frac{12+13}{2} = 12.5$
    The mode is 13.
    $s = \sqrt{\frac{(12.6-17)^2 + (12.6-8)^2 + \cdots + (12.6-12)^2}{30}} = \sqrt{\frac{381.2}{30}} \doteq 3.6$

11. (a) Q_L is the median of the first 15 ordered values, the 8th value.
    $Q_L = 10$
    Q_U is the median of the last 15 ordered values, the 23rd value.
    $Q_U = 15$

    (b) 5-number summary; minimum $-Q_L - \hat{x} - Q_U -$ maximum
    For Problem 1 data: $7 - 10 - 12.5 - 15 - 21$

    (c) Q_L is the median for the first 12 ordered values, the average of the 6th and 7th values.
    $Q_L = \frac{8+8}{2} = 8$
    Median is the average of the 12th and 13th ordered values.
    $\hat{x} = \frac{9+9}{2} = 9$
    Q_U is the median for the last 12 ordered values, the average of the 18th and 19th values.
    $Q_U = \frac{11+11}{2} = 11$

    (d) 5-number summary for the Problem 4 data: $6 - 8 - 9 - 11 - 13$

    (e)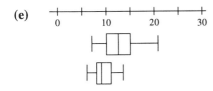

12. (a) $\bar{x} = \dfrac{1 \cdot 24 + 4 \cdot 25 + 7 \cdot 26 + 8 \cdot 27 + 6 \cdot 28 + 5 \cdot 29 + 5 \cdot 30 + 3 \cdot 31 + 1 \cdot 33}{1 + 4 + 7 + 8 + 6 + 5 + 5 + 3 + 1} = \dfrac{1111}{40} \doteq 27.8$

$s = \sqrt{\dfrac{1 \cdot (27.8 - 24)^2 + 4 \cdot (27.8 - 25)^2 + 7 \cdot (27.8 - 26)^2 + \cdots + 1 \cdot (27.8 - 33)^2}{40}} = \sqrt{\dfrac{163}{40}} \doteq 2.0$

(b) $\bar{x} = \dfrac{1 \cdot 24 + 3 \cdot 25 + 8 \cdot 26 + 12 \cdot 27 + 9 \cdot 28 + 4 \cdot 29 + 2 \cdot 30 + 1 \cdot 33}{1 + 3 + 8 + 12 + 9 + 4 + 2 + 1} = \dfrac{1092}{40} = 27.3$

$s = \sqrt{\dfrac{1 \cdot (27.3 - 24)^2 + 3 \cdot (27.3 - 25)^2 + 8 \cdot (27.3 - 26)^2 + \cdots + 1 \cdot (27.3 - 33)^2}{40}} = \sqrt{\dfrac{104.4}{40}} \doteq 1.6$

(c) The means are about the same for the two sets of data, but the standard deviation is smaller for the second histogram since the data is less spread out from the mean.

13. There are $21 \cdot 77 = 1617$ points for the 21 students. So there are 1839 points for all 24 students. Thus, the average is $1839 \div 24 \doteq 76.6$.

14. There are $27 \cdot 75 = 2025$ points for the second period students and $30 \cdot 78 = 2340$ points for the fourth period students. Thus, the average for all students is $\dfrac{2340 + 2025}{57} \doteq 76.6$.

15. No. The sample only represents the population of students at State University, not university students nationwide.

16. You might want to limit your sample to persons 20 years old and older who want to work. Alternatively, you might want to define several populations and determine figures for each: persons 20 years old and older who want to work, teenagers who want full-time employment, teenagers who want part-time employment, adults 20 years old and older who want part-time employment, and so on.

17. Telephone polls sample only persons sufficiently affluent to own a telephone. They are also biased by the fact that many people do not like telephone polls or telephone commercial solicitations and so refuse to respond or respond inaccurately because of anger.

18. Voluntary responses to mailed questionnaires tend to come primarily from those who feel strongly (either positively or negatively) about an issue or who represent narrow special interest groups. They are rarely representative of the population as a whole.

19. One way is to number the students in alphabetical order and select the sample using a spinner or a random number generator on a computer. Alternatively, one might print the names of all students on slips of paper, place them in a container, mix them well, and have someone close their eyes and select from the container the names of those to be in the sample.

20. Yes. Just continue taking samples until one finally shows up with eight out of the ten in the sample preferring WHITO toothpaste. As long as the population of dentists contains at least 8 dentists who prefer WHITO Toothpaste, eventually a sample will contain those 8 out of 10.

Chapter Test (page 547)

1. (a)
```
                                          X
                               X XX      XXX
     XX            X    X X X  XXXXX XX  XXXXXXX XX X
   ┬────┬────┬────┬────┬────┬────┬────┬
   40   50   60   70   80   90   100
```

(b) 41 and 42 are certainly outliers, since they are so far removed from the major collection of data.

2. (a) $\bar{x} = \dfrac{2356}{30} \doteq 78.5$

(b) The median is the average of the 15th and 16th ordered values.
$\hat{x} = \dfrac{81 + 81}{2} = 81$

(c) mode = 87

(d) $s \doteq \sqrt{\dfrac{(78.5-42)^2+(78.5-86)^2+\cdots+(78.5-57)^2}{30}} \doteq \sqrt{\dfrac{5217.5}{30}} \doteq 13.2$

3.
```
4 | 1  2
5 | 7
6 | 3  6
7 | 0  4  5  6  7  8  8
8 | 0  0  1  1  4  5  6  6  7  7  7  8  8  9
9 | 0  2  3  5
```

4. The 5-number summary for the data in Problem 1: 41 – 75 – 81 – 87 – 95

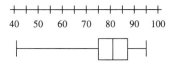

5. $(0.40) \cdot (360°) = 144°$

6. A random sample is one chosen in such a way that every subset of size r in the population has an equal chance of being included.

7. To average 80% on all five tests, Nanda must earn a total of 400 points. Nanda already has 77 + 79 + 72 = 228 points. Therefore, she must score at least 400 – 228 = 172 points on the remaining two exams.

8. With a normal distribution, 68% of the data will be within one standard deviation of the mean, 95% of the data will be within two standard deviations of the mean and 99.7% of the data will be within three standard deviations of the mean.

9. One way would be to number the students and select 200 using random numbers from a random number generator to determine which students are included in the sample.

10. If the sample is not chosen at random it is quite likely to reflect bias—bias of the sampler, bias reflecting the group from which the sample was actually chosen (views of teamsters, or AARP members), and so on.

Chapter 9

JUST FOR FUN   Three Card Monte (page 558)

(a) There are three possibilities for the three cards: BB, BW, and WW. If a card is dealt at random and a black side comes up, then the card is either the BB card of the BW card. Thus, it might appear that the probability that the other side is black is $\frac{1}{2}$.

However, the BB card can come up either way and still show B. Thus, in two out of three cases, a card showing B is B on the other side as well. Thus, the probability in question is $\frac{2}{3}$.

(b) We did 15 trials, obtaining 11 black on the second side for a probability of $P_e \doteq 0.73$. This tends to confirm the above guess.

Problem Set 9.1 (page 561)

1. 4 of the 35 scores are 79s. Therefore,
$P_e(79) = \frac{4}{35} \doteq 0.11$.

4. (b) Individual probabilities will vary but the sum should always be 1. In our case, we have,
$P_e(1) + P_e(2) + P_e(3) + P_e(4) + P_e(5)$
$= 0.12 + 0.22 + 0.22 + 0.22 + 0.22$
$= 1.00$.

7. (a) $P_e(A) = 0$ means that event A never happened in the simulation. This corresponds to choice iv (or possibly ii).

(c) $P_e(C) = -0.5$ means that event C happened a negative number of times which is not possible. This corresponds to choice ii.

(e) $P_e(E) = 1.7$ means that event E happened more than 100% of the time which is not possible. This corresponds to choice ii.

9. (b)

red die	+	green die
6	+	2
5	+	3
4	+	4
3	+	5
2	+	6

11. (a) Answers will vary. For the 20 spins, the line on the thumb ended up in Region A 12 times, Region B 6 times, and Region C only twice. Therefore,
$P_e(A) = \frac{12}{20} = 0.6$, $P_e(B) = \frac{6}{20} = 0.3$,
$P_e(C) = \frac{2}{20} = 0.1$.

(b) Answers will vary. For 20 spins, the line on the thumb should end up in Region A $\left(\frac{240}{360}\right) \cdot 20 \doteq 13$ times. This is close to the observed 12 times. The line on the thumb should end up in Region B $\left(\frac{90}{360}\right) \cdot 20 = 5$ times. This is close to the observed 6 times. The line on the thumb should end up in Region C $\left(\frac{30}{360}\right) \cdot 20 \doteq 2$ times. That is exactly the observed result.

13. Answers will vary. For the experiment with 10 trials, shown above, the first 6 occurred on the fourth roll twice. Therefore,
$P_e(\text{first 6 on fourth roll}) = \frac{2}{10} = 0.2$.

16. Answers will vary. Here is a solution that differs from the one shown in the back of the textbook. The following cards were selected:

Hearts	Diamonds	Spades	Clubs
6	5	K	8
9	2	8	7
Q	Q	J	5
2	J	10	9
	2	5	
	6		
	J		

(a) Empirical probability of a red card = $\dfrac{\text{diamonds + hearts}}{20}$

$P_e(R) = \dfrac{4+7}{20} = \dfrac{11}{20} = 0.55$

(b) Empirical probability of a face card = $\dfrac{J+Q+K}{20}$

$P_e(F) = \dfrac{3+2+1}{20} = \dfrac{6}{20} = 0.3$

(c) Empirical probability of a red card or a face card

$= \dfrac{\text{diamonds + hearts + spades(J or Q or K) + clubs(J or Q or K)}}{20}$

$P_e(R \text{ or } F) = \dfrac{4+7+2+0}{20} = \dfrac{13}{20} = 0.65$

(d) Empirical probability of a red card and a face card = $\dfrac{\text{diamonds(J or Q or K) + hearts(J or Q or K)}}{20}$

$P_e(R \text{ and } F) = \dfrac{1+3}{20} = \dfrac{4}{20} = 0.20$

(e) $P_e(R) + P_e(F) - P_e(R \text{ and } F) = 0.55 + 0.3 - 0.20 = 0.65$

(f) The empirical probability of a red card or a face card can be found using the expression in (e), that is, $P_e(R \text{ or } F) = P_e(R) + P_e(F) - P_e(R \text{ and } F)$. In words, sum the empirical probability of a red card with the empirical probability of a face card then subtract the empirical probability of a red face card.

(g) In part (e) of Problem 15, the empirical probability of rolling a 7 or 8 was found by simply summing the empirical probabilities of each event. In part (f) of Problem 16, the empirical probability of a red card or a face card was also found by summing the empirical probabilities of each event, but there was an additional step of subtracting the empirical probability of a red card and a face card. To understand this additional step, consider the queen of hearts, queen of diamonds, and the two jacks of diamonds that were drawn in the experiment. These 4 cards are "double counted" in the expression: $P_e(R) + P_e(F)$ because they are red *and* face cards.

$P_e(R) + P_e(F) = \dfrac{\text{diamonds + hearts}}{20} + \dfrac{J + Q + K}{20}$

To correct the "double counting" problem, these 4 cards must be subtracted. This is achieved by subtracting $P_e(R \text{ and } F)$. Therefore, $P_e(R \text{ or } F) = P_e(R) + P_e(F) - P_e(R \text{ and } F)$. The expression for part (e) in Problem 15 is similar and in fact could be written:

$P_e(7 \text{ or } 8) = P_e(7) + P_e(8) - P_e(7 \text{ and } 8)$.

Notice that $P_e(7 \text{ and } 8) = 0$ since the dice can never be a 7 *and* an 8.

18. Answers will vary. When we did the experiment we obtained these outcomes:
H5 H3 H2 H3 H4 T3 T2 T2 H2 T1 H2 H2 T4 T5 H1 H5 T1 H6 H4 T1

(a) $P_e(H) = \dfrac{12}{20} = 0.6$

(b) $P_e(5) = \dfrac{3}{20} = 0.15$

(c) $P_e(H \text{ and } 5) = \dfrac{2}{20} = 0.1$

(d) $P_e(H) \cdot P_e(5) = (0.6) \cdot (0.15) = 0.09$

(e) Yes. Since the events H and 5 are independent, the number of simultaneous occurrences of H and 5 should be about $P_e(H) \cdot$ (the number of occurrences of 5). But then

$P_e(H \text{ and } 5) \doteq \dfrac{P_e(H) \cdot (\text{the number of occurrences of 5})}{20}$

$= P_e(H) \cdot P_e(5)$.

24. Answers will vary. Using the points generated for the solution of Problem 23, we compute the following empirical probabilities.

(e) Area for (a): $\left(\frac{1}{2}\right) \cdot (0.5) \cdot (0.4) = 0.1$

Area for (b): Draw a horizontal line at 0.4 to form 2 triangles. $\left(\frac{1}{2}\right) \cdot (0.6) \cdot (0.2) + \left(\frac{1}{2}\right) \cdot (0.6) \cdot (0.2)$
$= 0.06 + 0.06 = 0.12$

Area for (c): This is a circle with radius
$r = 0.4$. $\pi r^2 = \pi \cdot (0.4)^2 \doteq 0.50$

The Monte Carlo method did not estimate the area for (a) and (b) very well. These are relatively small regions and 20 points is not enough to provide a thorough coverage of the entire unit square and identify the area of each figure. The Monte Carlo method did come quite close for the circle in (c). This is because the circle occupies about half of the unit square so 20 points was enough to cover the region and identify the area of the circle with about half the points inside the circle and half the points outside.

25. (a) Of the 9,664,994 people alive at 20 years old, only 2,626,372 will still be alive at age 80. Therefore,

P_e(a person 20 years old lives to be at least 80) $= \dfrac{2,626,372}{9,664,994} \doteq 0.27$.

30. $s = \sqrt{\dfrac{(30-30)^2 + (30-28)^2 + \cdots + (30-31)^2}{8}}$
$= \sqrt{\dfrac{44}{8}} \doteq 2.3$

32. Assuming that the population is normal, the probability that an individual falls within the given range is 95% since 11.2 and 44.8 are two standard deviations from the mean.

JUST FOR FUN   Social Security Numbers (page 566)

Answers will vary. On the average, over 90% of the numbers will have an even digit in the fifth position.

Problem Set 9.2

1. There are 3 ways to roll a 4:
   $4 = 1 + 3 = 2 + 2 = 3 + 1$.
   There are 5 ways to roll a 6:
   $6 = 1 + 5 = 2 + 4 = 3 + 3 = 4 + 2 = 5 + 1$.
   Therefore, there are $3 + 5 = 8$ ways to roll a 4 or 6.

3. The 12 possible outcomes are:
   H1 H2 H3 H4 H5 H6 T1 T2 T3 T4 T5 T6

   (a) There are 3 ways to get a head and an even number: H2 H4 H6

5. Draw a Venn Diagram with 2 overlapping circles. Label one circle "E" for English and the other circle "J" for Japanese. Since 11 of the students speak only English, the nonoverlapping part of "E" contains 11 members. 9 of the students speak only Japanese so the nonoverlapping part of "J" contains 9 members. That accounts for $11 + 9 = 20$ students out of 24 students in Mr. Walcott's class, so 4 students must be able to speak both English and Japanese (the overlapping part of "E" and "J").

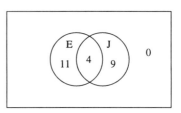

7. Let R be the set of red face cards.
   {jack of diamonds, queen of diamonds, king of diamonds, jack of hearts, queen of hearts, king of hearts}

   Let A be the set of black aces.
   {ace of spades, ace of clubs}

   Notice that $R \cap A = \varnothing$ since there are no elements that are common to both set R and set A. Since sets R and A are mutually exclusive, use the second principle of counting:
   $n(R \text{ or } A) = n(R) + n(A) = 6 + 2 = 8$.
   There are 8 ways to select a red face card or a black ace from an ordinary deck of playing cards.

8. (a) Since there are 6 numbers to choose from, the first digit has 6 possibilities. After the first digit is selected, the number cannot be repeated so there are 5 numbers to choose from for the second digit. The third digit has 4 remaining possibilities and the fourth digit has 3 numbers to choose from. Use the third principle of counting to find:

    n(four digit natural numbers with no repeat digits)
    $= n(\text{1st digit}) \cdot n(\text{2nd digit} | \text{1st digit}) \cdot n(\text{3rd digit} | \text{1st and 2nd digits})$
    $\cdot n(\text{4th digit} | \text{1st and 2nd and 3rd digits})$
    $= 6 \cdot 5 \cdot 4 \cdot 3 = 360$

    There are 360 four digit natural numbers using the digits 1, 2, 3, 4, 5, or 6 at most once.

    (b) 3 of the 6 digits are odd so there are 3 possibilities for the first digit. Once the first digit is chosen, there are 5 possibilities for the second digit, 4 possibilities for the third digit, and 3 possibilities for the fourth digit. Use the same reasoning presented in (a) to find that there are $3 \cdot 5 \cdot 4 \cdot 3 = 180$ four digit natural numbers using the digits 1, 2, 3, 4, 5, or 6 at most once and beginning with an odd digit.

    (c) 3 of the 6 digits are odd and one of the three digits must appear in the first position. That leaves 2 of the 3 odd digits for the last position. Once the first and last digits are chosen, there are 4 possibilities for the second digit and 3 possibilities for the third digit. Use the same reasoning presented in (a) to find that there are $3 \cdot 2 \cdot 4 \cdot 3 = 72$ four digit natural numbers using the digits 1, 2, 3, 4, 5, or 6 at most once and beginning and ending with an odd digit.

10. To draw the possibility tree, begin by showing the 3 possible letters for the first letter and label these 3 limbs "a", "b", and "c". Now there are two possible letters remaining for the second letter so draw 2 limbs from each of the first letters and label each limb with the remaining possible letters. There is only one choice for the third letter after the first two have been selected so draw a limb and label the remaining letter. The possibility tree is shown below.

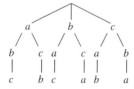

    Following the limbs of the possibility tree, the code words are: $abc, acb, bac, bca, cab, cba$.

12. (a) $7! = 7 \cdot 6 \cdot 5 \cdot 4 \cdot 3 \cdot 2 \cdot 1 = 5040$

    (c) $9! \div 7! = \dfrac{9 \cdot 8 \cdot 7 \cdot 6 \cdot 5 \cdot 4 \cdot 3 \cdot 2 \cdot 1}{7 \cdot 6 \cdot 5 \cdot 4 \cdot 3 \cdot 2 \cdot 1} = 9 \cdot 8 = 72$

    (e) $7 \cdot 7! = 7 \cdot 7 \cdot 6 \cdot 5 \cdot 4 \cdot 3 \cdot 2 \cdot 1 = 7 \cdot 5040 = 35{,}280$

13. (a) $P(13, 8) = 13(13-1)(13-2)(13-3)(13-4)(13-5)(13-6)(13-7)$
    $= 13 \cdot 12 \cdot 11 \cdot 10 \cdot 9 \cdot 8 \cdot 7 \cdot 6 = 51{,}891{,}840$

    (c) $P(15, 2) = 15(15-1) = 15 \cdot 14 = 210$

    (e) $C(15, 15) = \dfrac{15(15-1)(15-2)\cdots(15-13)(15-14)}{15!} = \dfrac{15 \cdot 14 \cdot 13 \cdot 12 \cdot 11 \cdot 10 \cdot 9 \cdot 8 \cdot 7 \cdot 6 \cdot 5 \cdot 4 \cdot 3 \cdot 2 \cdot 1}{15 \cdot 14 \cdot 13 \cdot 12 \cdot 11 \cdot 10 \cdot 9 \cdot 8 \cdot 7 \cdot 6 \cdot 5 \cdot 4 \cdot 3 \cdot 2 \cdot 1} = 1$

15. The digits 1, 3, 5, 7, and 9 are odd and the digits 0, 2, 4, 6, and 8 are even.

    (a) There are 5 choices for the first odd digit, 5 choices of the second odd digit, 5 choices for the third odd digit, 5 choices for the fourth even digit, and 5 choices for the last even digit. Therefore, there are $5 \cdot 5 \cdot 5 \cdot 5 \cdot 5 = 5^5 = 3125$ 5-digit numbers with the first three digits odd and the last two digits even when repetition of digits is allowed.

17. (a) There are 4 letters with two T's and two O's. The number of permutations of 4 things with 2 things alike and 2 other things alike is given by:
$$\frac{4!}{2!2!} = \frac{4\cdot 3\cdot 2\cdot 1}{2\cdot 1\cdot 2\cdot 1} = \frac{4\cdot 3}{2\cdot 1} = 6.$$
There are 6 different arrangements of the letters in TOOT.

18. From Example 9.12, the scores and the number of ways to obtain the score are listed in the table below.

score (sum)	2	3	4	5	6	7	8	9	10	11	12
# of ways	1	2	3	4	5	6	5	4	3	2	1

(a) The scores 2, 4, 6, 8, 10, and 12 are even and can be obtained $1 + 3 + 5 + 5 + 3 + 1 = 18$ ways.

21. (a) For every arrangement under consideration b immediately follows a so the sequence ab can be thought of as a single symbol. That is, how many arrangements of ab, c, d, and e are there? There are 4 choices for the first position, 3 choices for the second position, 2 choices for the third position, and one choice left for the last position. There are $4 \cdot 3 \cdot 2 \cdot 1 = 4! = 24$ of the 120 arrangements of a, b, c, d, and e where b immediately follows a.

(b) ab and ba are the two sequences where a and b are adjacent. Use the same reasoning as part (a) to find that there are 4! arrangements of ab, c, d, and e and 4! arrangements of ba, c, d, and e for a total of $4! + 4! = 24 + 24 = 48$ of the 120 arrangements of a, b, c, d, and e where a and b are adjacent.

(c) Consider all the cases where a precedes e.

(i) If a is first, e could be second, third, fourth, or fifth. This can happen as shown.
$1 \cdot 1 \cdot 3 \cdot 2 \cdot 1 = 6$
$1 \cdot 3 \cdot 1 \cdot 2 \cdot 1 = 6$
$1 \cdot 3 \cdot 2 \cdot 1 \cdot 1 = 6$
$1 \cdot 3 \cdot 2 \cdot 1 \cdot 1 = 6$
$\overline{}$
24 ways.

(ii) If a is second, e could be third, fourth, or fifth. This can happen as shown.
$3 \cdot 1 \cdot 1 \cdot 2 \cdot 1 = 6$
$3 \cdot 1 \cdot 2 \cdot 1 \cdot 1 = 6$
$3 \cdot 1 \cdot 2 \cdot 1 \cdot 1 = 6$
$\overline{}$
18 ways.

(iii) If a is third, e could be fourth or fifth. This can happen as shown.
$3 \cdot 2 \cdot 1 \cdot 1 \cdot 1 = 6$
$3 \cdot 2 \cdot 1 \cdot 1 \cdot 1 = 6$
$\overline{}$
12 ways.

(iv) If a is fourth, e could be fifth. This can happen in $3 \cdot 2 \cdot 1 \cdot 1 \cdot 1 = 6$ ways.

Combining the 4 cases above, there are $24 + 18 + 12 + 6 = 60$ ways a can precede e.
Alternatively, by symmetry, in half the $5! = 120$ permutations of a, b, c, d, and e, a precedes e and in half e precedes a. Therefore, the desired number is $\frac{120}{2} = 60$ as above.

26. (a) There are
$$C(13, 3) = \frac{13 \cdot 12 \cdot 11}{3 \cdot 2 \cdot 1} = \frac{1716}{6} = 286 \text{ ways}$$
to choose the three boys for the committee. There are
$$C(11, 3) = \frac{11 \cdot 10 \cdot 9}{3 \cdot 2 \cdot 1} = \frac{990}{6} = 165 \text{ ways}$$
to choose the three girls for the committee.
That means there are $C(13, 3) \cdot C(11, 3) = 286 \cdot 165 = 47{,}190$ ways to select a committee of three boys and three girls.

32. One play at roulette is independent of any other play at roulette because previous or future outcomes do not affect the current outcome. Francine's 20 previous losses will not affect the next outcome. Her "luck" is determined with each play, not with a string of independent trials. She should place her bet based on the odds of her strategy. If she is losing with every play, she should consider changing her strategy rather than bet heavily with the same losing strategy.

34. (a)

```
                                        x
                                   x    x
        x  xx  x   x  xx   x x  x  xx   x x x
       ┬──┬──┬──┬──┬──┬──┬──┬──┬──┬──┬──┬──┬
       40    50    60    70    80    90    100
```

(c) Smallest value is 53, largest value is 90, Q_L is the fifth ordered value so $Q_L = 64$, median is the average of the ninth and tenth ordered value so $\hat{x} = \frac{74+77}{2} = 75.5$, Q_u is the 14th ordered value so $Q_u = 82$. 5-number summary: 53 – 64 – 75.5 – 82 – 90.

JUST FOR FUN   It's a Girl! (page 585)

The possibilities in order of age are boy, boy; boy, girl; girl, boy; girl, girl. In three cases there is at least one boy and in two of these the other child is a girl. Thus, the desired probability is $\frac{2}{3}$. (The point here is that more 2-child families have a boy and a girl than have two boys. On the other hand, if one were to choose at random from the set of boys who have one sibling, the probability that the other sibling is a boy would indeed be $\frac{1}{2}$.)

JUST FOR FUN   A Probability Paradox (page 586)

For the three tables, the probabilities are as follows
Table A: $P(R|G) = \frac{5}{11} \doteq 0.45$, $P(R|B) = \frac{3}{7} \doteq 0.43$
Table B: $P(R|G) = \frac{6}{9} \doteq 0.67$, $P(R|B) = \frac{9}{14} \doteq 0.64$
Table C: $P(R|G) = \frac{11}{20} = 0.55$, $P(R|B) = \frac{12}{21} \doteq 0.57$

This is quite surprising since $P(R|G) > P(R|B)$ on each of tables A and B, but $P(R|G) < P(R|B)$ on table C where the balls in the hats were obtained by combining the balls in the hats on tables A and B. This is another example of Simpson's paradox (see page 411).

Problem Set 9.3

1. (a) There are two possible outcomes for each coin: Head (H) or Tail (T). The list of all outcomes for the experiment in order of penny, nickel, dime, and quarter are:

   HHHH THHH TTHH TTTH TTTT
         HTHH THTH TTHT
         HHTH THHT THTT
         HHHT HTTH HTTT
                HTHT
                HHTT

   (c) There are 16 possible outcomes for the experiment and six of the outcomes have two heads and two tails (center column of outcomes from (a)). Therefore,
   $P(2 \text{ heads and } 2 \text{ tails}) = \frac{6}{16} = 0.375$.

2. $P(3 \text{ or } 4) = P(3) + P(4) = \frac{2}{36} + \frac{3}{36}$
   $= \frac{5}{36} \doteq 0.14$

3. $S = \{3 \text{ red Fords, } 4 \text{ white Fords, } 2 \text{ black Fords, } 6 \text{ red Hondas, } 2 \text{ white Hondas, } 5 \text{ black Hondas}\}$
   There are 22 equally likely cars to randomly choose from.

   (a) Of all of the 22 cars available, four are white Fords. Therefore, $P(\text{white Ford}) = \frac{4}{22} \doteq 0.18$.

(c) There are 6 white cars (4 white Fords and 2 white Hondas) from the 22 available cars. Therefore, $P(\text{white}) = \frac{6}{22} \doteq 0.27$.

4. $S = \{B1, B2, B3, B4, B5, W1, W2, W3, W4, W5, W6, W7\}$
   There are 12 equally likely balls to randomly choose from.

   (a) Of the 12 balls to select, four are numbered 1 or 2 (B1, W1, B2, W2). Therefore,
   $P(1 \text{ or } 2) = P(1) + P(2) = \frac{2}{12} + \frac{2}{12} = \frac{4}{12} \doteq 0.33$.

5. Let BR = brown, BL = blue, br = brunette, bl = blond.
   $S = \{7\text{BRbr}, 2\text{BLbr}, 8\text{BLbl}, 3\text{BRbl}\}$. There are 20 equally likely children to randomly select.

   (a) There are 7 brown-eyed brunettes out of the 20 children in the class. Therefore,
   $P(\text{BRbr}) = \frac{7}{20} = 0.35$.

6. $S = \{00, 01, 02, \cdots, 98, 99\}$. There are 100 equally likely 2-digit numbers possible.

   (a) The set of numbers greater than 80 is $\{81, 82, \cdots, 98, 99\}$ and there are 19 such numbers. Therefore,
   $P(\text{a number greater than } 80) = \frac{19}{100} = 0.19$.

   (c) The set of numbers that are a multiple of 3 is $\{3, 6, 9, \cdots, 96, 99\}$ and there are $\frac{99}{3} = 33$ such numbers. Therefore, $P(\text{a number is a multiple of } 3) = \frac{33}{100} = 0.33$.

   (e) The set of even numbers less than 50 is $\{00, 02, \cdots, 46, 48\}$ and there are $\frac{50}{2} = 25$ such numbers. Therefore, $P(\text{an even number less than } 50) = \frac{25}{100} = 0.25$.

7. Assume the deck of cards is shuffled between deals so that the only known cards are the two dealt face up. That leaves 50 unknown cards.

   (a) In order to win, the third card must be a six. There are four sixes left in the deck. Therefore,
   $P(\text{win}) = \frac{4}{50} = 0.08$

   (c) In order to win, the third card must be between the two cards dealt face, in this case, between a nine and a nine. There are no cards in the deck between two nines, therefore, $P(\text{win}) = \frac{0}{50} = 0$.

9. 3 cards are randomly dealt from a deck of 52 cards in $C(52, 3)$ ways. 3 hearts can be randomly dealt in $C(13, 3)$ ways. Therefore,
$$P(3 \text{ hearts}) = \frac{\text{number of ways for 3 hearts}}{\text{number of ways for 3 cards}} = \frac{C(13, 3)}{C(52, 3)}$$
$$= \frac{\frac{13 \cdot 12 \cdot 11}{3 \cdot 2 \cdot 1}}{\frac{52 \cdot 51 \cdot 50}{3 \cdot 2 \cdot 1}} = \frac{13 \cdot 12 \cdot 11}{52 \cdot 51 \cdot 50} = \frac{1716}{132,600} \doteq 0.013.$$

11. The sections 1–10 have equal areas so the central angle for each sector is $\frac{360°}{10} = 36°$. The sections a–e have equal areas so the central angle for each sector is $\frac{360°}{5} = 72°$. The probability for each of the following solutions are determined by ratios of angular measures of appropriate regions.

    (a) Three of the 10 regions are shaded so $P(\text{shaded}) = \frac{3 \cdot 36°}{360°} = \frac{108°}{360°} = 0.3$.

(c) Regions 10 and 6 represent two of the 10 regions so
$P(\text{region 10 or region 6}) = P(\text{region 10}) + P(\text{region 6})$
$$= \frac{36°}{360°} + \frac{36°}{360°} = \frac{72°}{360°} = 0.2$$

(e) Region 8 is one of three shaded areas so $P(\text{Region 8}|\text{shaded area}) = \frac{36°}{3 \cdot 36°} = \frac{36°}{108°} \doteq 0.33$.

(g) The vowels a and e cover 4 regions: 7, 8, 9, and 10. The odd regions are 1, 3, 5, 7, and 9. Regions 7 and 9 are odd and contain a vowel. Therefore,
$P(\text{a vowel or odd numbered region}) = P(\text{a vowel}) + P(\text{odd region}) - P(\text{vowel and odd})$
$$= \frac{4 \cdot 36°}{360°} + \frac{5 \cdot 36°}{360°} - \frac{2 \cdot 36°}{360°} = \frac{144° + 180° - 72°}{360°} = \frac{252°}{360°} = 0.7.$$

12. Notice the dart board measures 5 units wide and 5 units tall. To determine the area for region 5, notice that it is a square 1 unit by 1 unit so region 5 covers 1 sq. unit. To determine the areas for regions 1 and 3, find the area of the outside square and subtract the inside square area. The area for region 1 is $5^2 - 3^2 = 16$ sq. units. The area for region 3 is $3^2 - 1^2 = 8$ sq. units. The probabilities found below are ratios of regional areas.

(a) Region 1 takes up 16 of the 25 sq. units so $P(1) = \frac{16}{25} = 0.64$.

(c) Region 5 takes up 1 of the 25 sq. units so $P(5) = \frac{1}{25} = 0.04$.

13. Area of a circle = πr^2
Area of the dart board = $\pi(2)^2 = 4\pi$
Area of the inner circle = $\pi(1)^2 = \pi$
Area of each of the regions 1 through 10 = $4\pi - \pi = 3\pi$
Area of each of the regions a through e = $\frac{\pi}{5}$
Since the dart hits the board at random locations, the probability of hitting any particular region is the ratio of the area of the region and the area of the dart board.

(a) The area of region b is $\frac{\pi}{5}$ and the area of the dart board is 4π. Therefore,
$$P(\text{region b}) = \frac{\frac{\pi}{5}}{4\pi} = \frac{\pi}{5} \cdot \frac{1}{4\pi} = \frac{1}{20} = 0.05.$$

(b) The area of region b is $\frac{\pi}{5}$ and the area of the inner circle is π. Therefore,
$$P(\text{not region b}|\text{inner circle}) = \frac{\pi - \frac{\pi}{5}}{\pi}$$
$$= \frac{\frac{4\pi}{5}}{\pi} = \frac{4\pi}{5} \cdot \frac{1}{\pi} = \frac{4}{5} = 0.8.$$

(c) Regions a, b, and c are of equal size. Therefore, $P(\text{b}|\text{a, b, or c}) = \frac{1}{3} \doteq 0.33$.

14. (a) There are 5 ways to roll a six so there must be $36 - 5 = 31$ ways to not roll a six. If 6 represents the event of rolling a score of six, then the odds in favor of 6 are $n(6)$ to $n(\overline{6})$ or 5 to 31 written 5:31.

16. Since A and C are mutually exclusive, $P(A \text{ or } C) = P(A) + P(C) = \frac{1}{2} + \frac{1}{6} = \frac{4}{6}$. Therefore, as in Example 9.32, the odds in favor of A or C are given by $\frac{P(A \text{ or } C)}{1 - P(A \text{ or } C)} = \frac{\frac{4}{6}}{1 - \frac{4}{6}} = \frac{4}{2} = \frac{2}{1}$ or 2:1.

18. To determine the probability for each roll, count the number of ways to roll the score and divide by 36.

roll	2	3	4	5	6	7	8	9	10	11	12
probability	$\frac{1}{36}$	$\frac{2}{36}$	$\frac{3}{36}$	$\frac{4}{36}$	$\frac{5}{36}$	$\frac{6}{36}$	$\frac{5}{36}$	$\frac{4}{36}$	$\frac{3}{36}$	$\frac{2}{36}$	$\frac{1}{36}$

Expected win $= 4\left(\frac{1}{36}\right) + 6\left(\frac{2}{36}\right) + 8\left(\frac{3}{36}\right) + \cdots + 6\left(\frac{2}{36}\right) + 4\left(\frac{1}{36}\right) = \frac{600}{36} \doteq 16.67$

You can expect to win about $16.67 per roll on this game.

19. (a) There are $C(6, 2)$ ways to choose 2 white balls and $C(6 + 8, 2) = C(14, 2)$ ways to choose any 2 balls from the urn. Therefore,

$$P(2 \text{ white}) = \frac{n(\text{ways to draw 2 white balls})}{n(S)} = \frac{C(6, 2)}{C(14, 2)}$$

$$= \frac{\frac{6 \cdot 5}{2 \cdot 1}}{\frac{14 \cdot 13}{2 \cdot 1}} = \frac{6 \cdot 5}{14 \cdot 13} = \frac{30}{182} \doteq 0.16.$$

20. (a) There are $C(8, 4)$ ways to draw 4 red balls and $C(8 + 5 + 6, 4) = C(19, 4)$ ways to draw any 4 balls from the urn. Therefore,

$$P(\text{all four red}) = \frac{n(\text{ways to draw 4 red balls})}{n(S)} = \frac{C(8, 4)}{C(19, 4)}$$

$$= \frac{\frac{8 \cdot 7 \cdot 6 \cdot 5}{4 \cdot 3 \cdot 2 \cdot 1}}{\frac{19 \cdot 18 \cdot 17 \cdot 16}{4 \cdot 3 \cdot 2 \cdot 1}} = \frac{8 \cdot 7 \cdot 6 \cdot 5}{19 \cdot 18 \cdot 17 \cdot 16} = \frac{1680}{93,024} \doteq 0.02.$$

21. There are $P(26, 5) = 26 \cdot 25 \cdot 24 \cdot 23 \cdot 22 = 7,893,600$ 5-letter code words without repetition of letters.

(a) If a code word begins with the letter a, then there are 25 choices for the second letter, 24 choices for the third letter, 23 choices for the fourth letter, and 22 choices for the last letter. There are $P(25, 4) = 25 \cdot 24 \cdot 23 \cdot 22 = 303,600$ ways to arrange the last 4 letters. Therefore,

$$P(\text{a code word begins with a}) = \frac{P(25, 4)}{P(26, 5)} = \frac{303,600}{7,893,600} \doteq 0.04.$$

24. There are $C(52, 5) = \frac{52 \cdot 51 \cdot 50 \cdot 49 \cdot 48}{5 \cdot 4 \cdot 3 \cdot 2 \cdot 1} = \frac{311,875,200}{120} = 2,598,960$ possible 5-card hands.

(a) There are $C(4, 2) \cdot C(48, 3) = \frac{4 \cdot 3}{2 \cdot 1} \cdot \frac{48 \cdot 47 \cdot 46}{3 \cdot 2 \cdot 1} = 103,776$ possible 5-card hands with exactly two aces.

Therefore, $P(\text{a five card hand has exactly two aces}) = \frac{C(4, 2) \cdot C(48, 3)}{C(52, 5)} = \frac{103,776}{2,598,960} \doteq 0.040.$

25. One method of solving this problem is given in the back of the textbook. There is another method. There are $6 \cdot 6 \cdot 6 \cdot 6 \cdot 6 \cdot 6 \cdot 6 = 6^7 = 279,936$ possible outcomes for tossing seven dice. If each number 1–6 is to appear at least once, then one of the numbers 1 through 6 must appear twice. For each such number, there are $\frac{7!}{2!} = \frac{7 \cdot 6 \cdot 5 \cdot 4 \cdot 3 \cdot 2 \cdot 1}{2 \cdot 1} = 2520$ possible outcomes with the chosen number appearing twice and each other number appearing once. There are six possible numbers that can appear twice. Therefore,

$$P(\text{every number appears}) = \frac{6 \cdot 2520}{279,936} = \frac{15,120}{279,936} \doteq 0.05.$$

26. (a) $P(A \text{ and } B) = P(B \text{ and } A)$
$P(A \text{ and } B) = P(A)P(B|A)$ $\Big\}$ $P(A)P(B|A) = P(B)P(A|B)$
$P(B \text{ and } A) = P(B)P(A|B)$

Since $P(A)P(B|A) = P(B)P(A|B)$, then $P(B|A) = \frac{P(B)P(A|B)}{P(A)}.$

(c) From part (a), $P(B|A) = \dfrac{P(B)P(A|B)}{P(A)}$.

Replacing $P(A)$ by its equal from part (b), we obtain $P(B|A) = \dfrac{P(B)P(A|B)}{P(B)P(A|B) + P(\overline{B})P(A|\overline{B})}$.

29. There are $C(7, 4) = \dfrac{7 \cdot 6 \cdot 5 \cdot 4}{4 \cdot 3 \cdot 2 \cdot 1} = \dfrac{840}{24} = 35$ ways to choose four yellow marbles and
$C(8, 4) = \dfrac{8 \cdot 7 \cdot 6 \cdot 5}{4 \cdot 3 \cdot 2 \cdot 1} = \dfrac{1680}{24} = 70$ ways to choose four blue marbles. Therefore, there are
$C(7, 4) + C(8, 4) = 35 + 70 = 105$ ways to choose four marbles of the same color.

31. If the 5-letter code word begins with a vowel, there are 5 possible letters for the first letter, 26 − 5 = 21 possible letters for the second letter, 5 − 1 = 4 possible letters for the third letter, 21 − 1 = 20 possible letters for the fourth letter, and 4 − 1 = 3 possible letters for the last letter. Therefore, there are 5 · 21 · 4 · 20 · 3 = 25,200 possible code words beginning with a vowel. Alternatively, if the 5-letter code word begins with a consonant, there are 21 · 5 · 20 · 4 · 19 = 159,600 possible code words beginning with a consonant. There are a total of 25,200 + 159,600 = 184,800 possible 5-letter code words without repetition of letters where vowels and consonants alternate.

CLASSIC CONUNDRUM Should the Contestant Switch? (page 601)

If the contestant picks the "rich" door the host reveals the "lesser" or "joke" door. If the contestant picks the "lesser" door the host reveals the "joke" door leaving the "rich" door closed. If the contestant picks the "joke" door, the host reveals the "lesser" door leaving the "rich" door. Thus, in two of the three cases the door selected by the contestant is not the "rich" door. The contestant should switch.

Chapter Review Exercises

1. Answers will vary. When we did the experiment we obtained the following:

4H and 0T	3H and 1T	2H and 2T	1H and 3T	0H and 4T
	ЖН	ЖН IIII	ЖН I	

The empirical probability of obtaining three heads and one tail for our experiment is:
$P_e(\text{3H and 1T}) = \dfrac{5}{20} = 0.25$.

2. Answers will vary. When we did the experiment we obtained the following:

2	3	4	5	6	7	8	9	10	11	12
I	I	II	II	III	III	IIII	I	II	I	

(a) The empirical probability of obtaining a 3 or a 4 for our experiment is:
$P_e(3 \text{ or } 4) = P_e(3) + P_e(4) = \dfrac{1}{20} + \dfrac{2}{20} = \dfrac{3}{20} = 0.15$.

(b) The empirical probability of obtaining a score of at least 5 for our experiment is:
$P_e(\text{at least } 5) = P_e(5) + P_e(6) + P_e(7) + P_e(8) + P_e(9) + P_e(10) + P_e(11) + P_e(12)$
$= \dfrac{2}{20} + \dfrac{3}{20} + \dfrac{3}{20} + \dfrac{4}{20} + \dfrac{1}{20} + \dfrac{2}{20} + \dfrac{1}{20} + \dfrac{0}{20} = \dfrac{16}{20} = 0.80$.

3. Answers will vary. We use the results from our experiment shown in Problem 2. Given that a score from 5 through 9 occurred, the empirical probability of obtaining a score from 5 through 7 for our experiment is: $P_e(5 \text{ or } 6 \text{ or } 7 | 5 \text{ or } 6 \text{ or } 7 \text{ or } 8 \text{ or } 9) = \dfrac{n(5 \text{ or } 6 \text{ or } 7)}{n(5 \text{ or } 6 \text{ or } 7 \text{ or } 8 \text{ or } 9)} = \dfrac{8}{13} \doteq 0.62$.

4. Answers will vary. In our study we obtained the following:

Chocolate	Other
⊬H⊬ II	⊬H⊬ ⊬H⊬ III

The empirical probability that chocolate is the favorite flavor of ice cream for our survey is:
$P_e(\text{chocolate}) = \frac{7}{20} = 0.35$.

5. Answers will vary. When we did the experiment we obtained the following:

Number of tacks landing point up	5	4	3	2	1	0
Number of trials	II	IIII	⊬H⊬ I	⊬H⊬ I	I	I

(a) The empirical probability that precisely three of the tacks land point up for our experiment is: $P_e(3) = \frac{6}{20} = 0.30$.

(b) The empirical probability that two or three of the tacks land point up for our experiment is:
$P_e(2 \text{ or } 3) = P_e(2) + P_e(3) = \frac{6}{20} + \frac{6}{20} = \frac{12}{20} = 0.60$.

6. Answers will vary. When we did the experiment we obtained the following:

number of rolls to get a 5 or 6	1	2	3	4	5	6	7
number of trials			II	IIIII	III		I

(a) If we choose the average or mean number of rolls for our estimate of the number of rolls required to obtain a 5 or a 6, we obtain $\bar{x} = \frac{3+3+4+4+4+4+5+5+5+7}{10} = 4.4$. Since the number of rolls is a whole number, we estimate that it will take about 4 rolls.

(b) The empirical probability that it takes precisely five rolls to obtain a 5 or 6 for the first time for our experiment is: $P_e(5 \text{ rolls to obtain a 5 or 6}) = \frac{3}{10} = 0.30$.

7. Answers will vary. When we did the experiment we obtained the following:

	Heart	Nonheart
Ace	I	II
Nonace	⊬H⊬ III	⊬H⊬ IIII

(a) The empirical probability of drawing an ace or a heart for our experiment is:
$P_e(\text{ace or heart}) = P_e(\text{ace}) + P_e(\text{heart}) - P_e(\text{ace and heart})$
$= \frac{3}{20} + \frac{9}{20} - \frac{1}{20} = \frac{11}{20} = 0.55$.

(b) The empirical probability of drawing the ace of hearts for our experiment is:
$P_e(\text{ace and heart}) = \frac{1}{20} = 0.05$.

(c) Given that the card drawn was a heart, the empirical probability of drawing the ace for our experiment is:
$P_e(\text{ace} | \text{heart}) = \frac{1}{9} \doteq 0.11$.

8. Reading Table 9.1, there are 9,664,994 20-year olds and 468,174 90-year olds, so the empirical probability that a 20-year old will live to be 90 years old is: $P_e(90 \text{ year old} | 20 \text{ year old}) = \frac{468,174}{9,664,994} \doteq 0.05$.

9. (a) There are two choices, head (H) or tail (T), for each coin. The possible outcomes for the three coins are:
   HHH   HHT   HTT   TTT
          HTH   THT
          THH   TTH

   (b) Use the orderly list in (a) to see there are $\frac{3!}{2!1!} = 3$ ways to obtain two heads and one tail.

10. (a) There are $C(15, 9) = \frac{15 \cdot 14 \cdot 13 \cdot 12 \cdot 11 \cdot 10 \cdot 9 \cdot 8 \cdot 7}{9 \cdot 8 \cdot 7 \cdot 6 \cdot 5 \cdot 4 \cdot 3 \cdot 2 \cdot 1} = \frac{1,816,214,400}{362,880} = 5005$ ways to choose nine players from among 15 players without regard to the position of the player.

   (b) There are $C(2, 1) = \frac{2}{1} = 2$ ways to select a pitcher and $C(3, 1) = \frac{3}{1} = 3$ ways to select a catcher. Once the pitcher and catcher have been selected, there are 13 players remaining to fill the other seven positions. The rest of the team can be selected in $C(13, 7) = \frac{13 \cdot 12 \cdot 11 \cdot 10 \cdot 9 \cdot 8 \cdot 7}{7 \cdot 6 \cdot 5 \cdot 4 \cdot 3 \cdot 2 \cdot 1} = \frac{8,648,640}{5040} = 1716$ ways.
   Therefore, the team can be selected in
   $2 \cdot 3 \cdot 1716 = 10,296$ ways.

11. Draw a Venn Diagram with three overlapping circles and label one circle "F" for French, another circle "E" for English, and the last circle "G" for German. 17 students speak German, French, and English so label the section where all three circles overlap 17. 24 students speak both French and English so the entire overlap between circles F and E must sum to 24. 17 of these students have been recognized so the other overlapping section must be for 24 – 17 = 7 students. 27 students speak German and English with 17 already recognized so the other overlapping section of circles G and E must be for 27 – 17 = 10 students. A total of 38 students speak English so the remaining section of circle E must be for
   38 – (17 + 7 + 10) = 4 students. The information provided recognizes the language categories for 38 of the 90 students in Ferry Hall. Since all of the 90 students belong somewhere inside the circles F, E, and G, the three remaining sections (French only, German only, and French and German only) must sum to 90 – 38 = 52.

   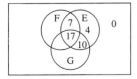

   (a) Use the Venn Diagram above and notice that there are only four students who do not speak French or German. Therefore, 90 – 4 = 86 students speak German or French.

   (b) Use the Venn Diagram above to find the overlapping section of circles F and E that does not overlap with circle G. There are 7 students who speak French and English but not German.

12. (a) There are 13 clubs in an ordinary deck of playing cards so there are
   $C(13, 2) = \frac{13 \cdot 12}{2 \cdot 1} = \frac{156}{2} = 78$ ways to select two clubs.

   (b) The face cards are jacks, queens, and kings and there are 12 all told in an ordinary deck of playing cards. Therefore, there are
   $C(12, 2) = \frac{12 \cdot 11}{2 \cdot 1} = \frac{132}{2} = 66$ ways to select two face cards.

   (c) Use $n(A \text{ or } B) = n(A) + n(B) - n(A \cap B)$ where event A is to select two clubs and event B is to select two face cards. From (a), $n(A) = 78$ and from (b), $n(B) = 66$. There are 3 face cards that are clubs so there are $C(3, 2) = \frac{3 \cdot 2}{2 \cdot 1} = 3$ ways to choose two face cards that are clubs.

Therefore,
$n(A \text{ or } B) = 78 + 66 - 3 = 141$. There are 141 ways to select two clubs or two face cards from an ordinary deck of playing cards.

13. (a) If repetition of letters is not allowed, each successive letter after the first has one less possible choice. That is, there are
$P(26, 5) = 26 \cdot 25 \cdot 24 \cdot 23 \cdot 22$
$= 7,893,600$ possible 5-letter words.

(b) There are 5 choices (a, e, i, o, u) for the first letter and 4 choices for the last letter. Removing these two letters leaves 24, 23, and 22 letters for the second, third, and fourth letters. Therefore, there are $5 \cdot 24 \cdot 23 \cdot 22 \cdot 4 = 242,880$ possible 5-letter code words that begin and end with vowels.

(c) The three letter sequence can appear in $C(3, 1) = \frac{3}{1} = 3$ sequence positions: aef _ _, _ aef _, or _ _ aef. There are $26 - 3 = 23$ letters remaining for the first blank and 22 letters remaining for the second blank. Therefore, there are $3 \cdot 23 \cdot 22 = 1518$ possible 5-letter code words which contain the sequence aef.

14. (a) STREETS has two S's, two T's, one R, and two E's. These letters can be arranged in
$\frac{7!}{2!2!1!2!} = \frac{7 \cdot 6 \cdot 5 \cdot 4 \cdot 3 \cdot 2 \cdot 1}{2 \cdot 1 \cdot 2 \cdot 1 \cdot 1 \cdot 2 \cdot 1} = \frac{5040}{8} = 630$ recognizably different orders.

(b) Now, STREETS has two S's, two T's, one R, and *one* EE. These letters can be arranged in
$\frac{6!}{2!2!1!1!} = \frac{6 \cdot 5 \cdot 4 \cdot 3 \cdot 2 \cdot 1}{2 \cdot 1 \cdot 2 \cdot 1 \cdot 1} = \frac{720}{4} = 180$ recognizably different orders with two E's adjacent.

(c) Since one T starts the word, there are two S's, one T, one R, and two E's for the remaining positions in the word. These letters can be arranged in $\frac{6!}{2!1!1!2!} = \frac{6 \cdot 5 \cdot 4 \cdot 3 \cdot 2 \cdot 1}{2 \cdot 1 \cdot 1 \cdot 2 \cdot 1} = \frac{720}{4} = 180$ recognizably different orders.

15. (a) There are two possible outcomes for each of the two coins, head (H) or tail (T), and six possible outcomes for the die (1, 2, 3, 4, 5, 6).
$$S = \begin{Bmatrix} HH1 & HH2 & HH3 & HH4 & HH5 & HH6 \\ HT1 & HT2 & HT3 & HT4 & HT5 & HT6 \\ TH1 & TH2 & TH3 & TH4 & TH5 & TH6 \\ TT1 & TT2 & TT3 & TT4 & TT5 & TT6 \end{Bmatrix}$$

(b) From (a), there are 24 equally likely outcomes and TT5 is one of those outcomes. Therefore, $P(TT5) = \frac{1}{24} \doteq 0.04$.

16. Use the sample space in 15(a) to note six outcomes that have tails for both coins. One of those six outcomes has a five for the die. Therefore, $P(5|TT) = \frac{1}{6} \doteq 0.17$.

17. Use Example 9.12 to obtain the following table:

Sum	2	3	4	5	6	7	8	9	10	11	12
Probability	$\frac{1}{36}$	$\frac{2}{36}$	$\frac{3}{36}$	$\frac{4}{36}$	$\frac{5}{36}$	$\frac{6}{36}$	$\frac{5}{36}$	$\frac{4}{36}$	$\frac{3}{36}$	$\frac{2}{36}$	$\frac{1}{36}$

Solution 1: To compute the probability of obtaining a sum of at most 11, add the probabilities associated with sums of 2 through 11.
$P(\text{sum at most } 11) = \frac{1}{36} + \frac{2}{36} + \frac{3}{36} + \frac{4}{36} + \frac{5}{36} + \frac{6}{36} + \frac{5}{36} + \frac{4}{36} + \frac{3}{36} + \frac{2}{36} = \frac{35}{36} \doteq 0.97$

Solution 2: If you fail to obtain a sum of at most 11 you obtain a sum of 12.
Use $P(A) = 1 - P(\overline{A})$ where A is a sum of at most 11. Then,
$P(\text{sum at most } 11) = 1 - P(\text{sum of } 12) = 1 - \frac{1}{36} = \frac{35}{36} \doteq 0.97$.

18. $S = \{5 \text{ white}, 6 \text{ red}, 4 \text{ black}\}$
    There are $C(5 + 6 + 4, 2) = C(15, 2) = \frac{15 \cdot 14}{2 \cdot 1} = \frac{210}{2} = 105$ ways to randomly select two balls.

    (a) There are $C(5, 2) = \frac{5 \cdot 4}{2 \cdot 1} = \frac{20}{2} = 10$ ways to select two white balls, $C(6, 2) = \frac{6 \cdot 5}{2 \cdot 1} = \frac{30}{2} = 15$ ways to select two red balls, and $C(4, 2) = \frac{4 \cdot 3}{2 \cdot 1} = \frac{12}{2} = 6$ ways to select two black balls. Therefore,
    $P(2 \text{ balls the same color}) = \frac{C(5, 2) + C(6, 2) + C(4, 2)}{C(15, 2)} = \frac{10 + 15 + 6}{105} = \frac{31}{105} \doteq 0.30$.

    (b) From (a), there are $C(5, 2) = 10$ ways to select two white balls, thus
    $P(2 \text{ white}) = \frac{C(5, 2)}{C(15, 2)} = \frac{10}{105} \doteq 0.10$.

    (c) From (a), there are $C(5, 2) + C(6, 2) + C(4, 2) = 10 + 15 + 6 = 31$ ways to select two balls the same color, and $C(5, 2) = 10$ ways to select two white balls. Therefore,
    $P(2 \text{ white} | 2 \text{ same color}) = \frac{C(5, 2)}{C(5, 2) + C(6, 2) + C(4, 2)} = \frac{10}{31} \doteq 0.32$.

19. There are 12 equal size outer sections, each with central angle $\frac{360°}{12} = 30°$. There are 4 equal sized inner sections, each with central angle $\frac{360°}{4} = 90°$. The probabilities below are found using the appropriate ratios of angular measures.

    (a) Sections b and 8 do not overlap so $b \cap 8 = \emptyset$. Therefore, $P(b \text{ and } 8) = 0$.

    (b) Since b and 8 do not overlap, they are mutually exclusive. Sum the probability for each event to find:
    $P(b \text{ or } 8) = P(b) + P(8)$
    $= \frac{90°}{360°} + \frac{30°}{360°} = \frac{120°}{360°} \doteq 0.33$.

    (c) Given that the spinner lands in section 8, it cannot land in section b. Therefore, $P(b|8) = 0$.

    (d) Section 2 is a proper subset of section b. Therefore,
    $P(b \text{ and } 2) = P(2) = \frac{30°}{360°} \doteq 0.08$.

    (e) Sections b and 2 are not mutually exclusive, therefore,
    $P(b \text{ or } 2) = P(b) + P(2) - P(b \text{ and } 2)$
    $= \frac{90°}{360°} + \frac{30°}{360°} - \frac{30°}{360°} = \frac{90°}{360°}$
    $= 0.25$.

    (f) Given that the spinner lands in section b, it can land on Sections 1, 2, or 3. Therefore, use (d) to find:
    $P(2|b) = \frac{P(2 \text{ and } b)}{P(b)} = \frac{\frac{30°}{360°}}{\frac{90°}{360°}} = \frac{30°}{90°}$
    $\doteq 0.33$.

20. The possible outcomes for tossing three coins are:
    HHH   HHT   HTT   TTT
          HTH   THT
          THH   TTH
    Each of these eight possible outcomes are equally likely.

    (a) Use the list of outcomes above to note there are three ways to get two heads and one tail so there are five ways to not get two heads and one tail. The odds in favor of two heads and one tail are:
    $n(2\text{H and 1T})$ to $n(\overline{2\text{H and 1T}})$ or $3:5$.

(b) There is only one outcome of three heads so there are seven outcomes that are not three heads. The odds in favor of three heads are: $N(HHH)$ to $n(\overline{HHH})$ or $1:7$.

21. (a) Using the method of Example 9.2, the odds in favor of 4 are given by the ratio
$$\frac{P(A)}{1-P(A)} = \frac{0.85}{1-0.85} = \frac{0.85}{0.15} = \frac{17}{3}$$ which is often written as $17:3$.

(b) Use $n(S) = n(A) + n(\overline{A}) = 17 + 8 = 25$; then $P(A) = \frac{n(A)}{n(S)} = \frac{17}{25} = 0.68$.

22. (a) Expected value
$= \$0(0.15) + \$5(0.50) + \$10(0.25) + \$20(0.10)$
$= \$2.50 + \$2.50 + \$2.00 = \$7.00$
The expected value of the game is $7.00.

(b) You expect to collect $7.00 per play but you must pay $10.00 per play for an average loss of $3.00 over the long run. If you want to make money, it is not wise to play. If your enjoyment during the play of the game is worth at least $3.00 per play, then you can play for fun.

23. Since the gender of children is independent, $P(\text{other two children are boys}) = \frac{1}{2} \cdot \frac{1}{2} = 0.25$.

This problem is likely to stir up debate among students who have read the Just for Fun exercise on page 585, so we also give a more detailed analysis. If the children are arranged by age, there are eight possible arrangements of three children where G represents girl and B represents boy.

```
GGG   GGB   GBB   BBB
      GBG   BGB
      BGG   BBG
```

The census taker is greeted by a girl so the possibility of BBB is eliminated. It may be tempting to reason as follows:
Since the census taker has no way of knowing the girl's age in relation to her two siblings, there are seven possibilities to consider. Of the seven possibilities, three have two boys and one girl. Therefore, the probability of the other two children being boys is:
$P(2 \text{ boys and } 1 \text{ girl} | \text{at least } 1 \text{ girl}) = \frac{3}{7}$
$\doteq 0.43$.

This reasoning is flawed because the simple fact that a girl answered the door means that the seven possibilities discussed above are *not* equally likely, since a girl is far more likely to answer the door in a GGG family than in a BBG family. However, the child who answers the door *is* equally likely to be youngest, middle, or oldest. If she is youngest, there are four possibilities—GGG, GGB, GBG, and GBB—and these are equally likely, so the probability that both siblings are boys is $\frac{1}{4}$.

The same argument holds if she is oldest or is neither youngest nor oldest.

Chapter Test (page 604)

1. Prepare a card as shown and ask a number of people to choose number. Calculate the empirical probability of choosing 3 as the number of times 3 is chosen divided by the number of people questioned.

2. This would be an empirical probability obtained by keeping records for a large number of trials of treating strep throat with penicillin.

3. **(a)** $7! = 7 \cdot 6 \cdot 5 \cdot 4 \cdot 3 \cdot 2 \cdot 1 = 5040$

   (b) $\frac{9!}{6!} = \frac{9 \cdot 8 \cdot 7 \cdot 6!}{6!} = 9 \cdot 8 \cdot 7 = 504$

   (c) $\frac{8!}{(8-8)!} = \frac{8!}{0!} = \frac{8 \cdot 7 \cdot 6 \cdot 5 \cdot 4 \cdot 3 \cdot 2 \cdot 1}{1}$
   $= 40{,}320$

   (d) $7 \cdot 6! = 7! = 5040$

   (e) $P(8, 5) = 8 \cdot 7 \cdot 6 \cdot 5 \cdot 4 = 6720$

   (f) $P(8, 8) = 8 \cdot 7 \cdot 6 \cdot 5 \cdot 4 \cdot 3 \cdot 2 \cdot 1 = 8!$ Use (c) $8! = 40{,}320$.

   (g) $C(9, 3) = \frac{9 \cdot 8 \cdot 7}{3 \cdot 2 \cdot 1} = \frac{504}{6} = 84$

   (h) $C(9, 9) = \frac{9 \cdot 8 \cdot 7 \cdot 6 \cdot 5 \cdot 4 \cdot 3 \cdot 2 \cdot 1}{9 \cdot 8 \cdot 7 \cdot 6 \cdot 5 \cdot 4 \cdot 3 \cdot 2 \cdot 1} = \frac{9!}{9!} = 1$

4. Draw a Venn Diagram of three overlapping circles. Label one circle "F" for French, another circle "G" for German, and the remaining circle "C" for Chinese. One student studies all three languages so write "1" in the section where all three circles overlap. Two students study French and Chinese so write $2 - 1 = $ "1" in the empty section of the overlapping part of circles F and C. Three students study German and Chinese so write $3 - 1 = $ "2" in the empty section of the overlapping part of circles G and C. Twelve students study German and French so write

12 – 1 = "11" in the empty section of the overlapping part of circles G and F. 17 students study Chinese so write 17 – (1 + 1 + 2) = "13" in the empty section of circle C. 29 students study German so write 29 – (11 + 1 + 2) = "15" in the empty section of circle G. 27 students study French so write 27 – (11 + 1 + 1) = "14" in the empty section of circle F. All of Mrs. Spangler's calculus students study at least one foreign language so there are no students outside the circles in the Venn Diagram.

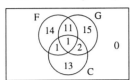

(a) To find the number of students in Mrs. Spangler's class, add the numbers in the circles on the Venn Diagram.
14 + 11 + 15 + 1 + 1 + 2 + 13 = 57

(b) Find the nonoverlapping section of circle C. 13 students only study Chinese.

(c) Find the overlapping section of circles F and G that does not include circle C. 11 students study French and German but not Chinese.

5. $S = \{Y_1, Y_2, Y_3, Y_4, Y_5, B_1, B_2, B_3, B_4, G_1, G_2, G_3, G_4, G_5, G_6, G_7, G_8\}$

(a) There are $C(8, 5) = \frac{8 \cdot 7 \cdot 6 \cdot 5 \cdot 4}{5 \cdot 4 \cdot 3 \cdot 2 \cdot 1} = \frac{6720}{120} = 56$ ways to select five green marbles.

(b) By the third principle of counting $n(A \text{ and } B) = n(A) \cap (B|A)$. If A and B are independent, $n(B|A) = n(B)$. If A and B are the events of choosing five yellow balls and five green balls, the events are independent and the desired number is $C(5, 5) \cdot C(8, 5) = 1 \cdot \frac{8 \cdot 7 \cdot 6}{1 \cdot 2 \cdot 3} = 56$.

(c) From (a), there are 56 ways to select five green marbles. There is only $C(5, 5) = \frac{5 \cdot 4 \cdot 3 \cdot 2 \cdot 1}{5 \cdot 4 \cdot 3 \cdot 2 \cdot 1} = \frac{5!}{5!} = 1$ way to select five yellow marbles. Therefore, there are 56 + 1 = 57 ways to select five yellow or five green marbles.

6. Since we are given that 3 of the balls selected are green, we must select 2 more of the 14 balls in $C(14, 2)$ ways. But we can select 2 yellow balls in $C(5, 2)$ ways. Thus,
$P(2 \text{ yellow}|3 \text{ green}) = \frac{C(5, 2)}{C(14, 2)} = \frac{\frac{5 \cdot 4}{1 \cdot 2}}{\frac{14 \cdot 13}{1 \cdot 2}} \doteq 0.11$.

7. There are five yellow marbles and 4 + 8 = 12 nonyellow marbles. The odds in favor of selecting a yellow marble is: $n(\text{yellow})$ to $n(\overline{\text{yellow}})$ or 5:12.

8. The odds in favor of E are: $\frac{P(E)}{1 - P(E)} = \frac{0.35}{1 - 0.35} = \frac{0.35}{0.65} = \frac{7}{13}$ or 7:13.

9. (a) There are seven choices for each letter so $7 \cdot 7 \cdot 7 \cdot 7 = 7^4 = 2401$ 4-letter code words are possible when repetition of letters is allowed.

(b) There are $P(7, 4) = 7 \cdot 6 \cdot 5 \cdot 4 = 840$ 4-letter code words when repetition of letters is not allowed.

10. (a) There are two vowels available to begin the word so the first letter has two possible choices. That leaves $P(6, 3) = 6 \cdot 5 \cdot 4 = 120$ ways to choose the remaining three letters. Therefore, there are $2 \cdot 120 = 240$ 4-letter words beginning with a vowel when repetition of letters is not allowed.

(b) Think of cd as a single point. There are 3 ways to position cd in a four letter code word and the other two positions can be filled in $5 \cdot 4 = 20$ ways. Therefore, there are $3 \cdot 20 = 60$ four letter code words with c followed by d. Similarly, there are 60 such words with d followed by c. Therefore, there are 60 + 60 = 120 four letter code words with c and d adjacent.

Chapter 10

JUST FOR FUN   Arranging Points at Integer Distances (page 613)

The isosceles trapezoid shown has bases of length 3" and 4". Its legs are 2" long and its diagonals are 4" long.

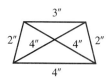

Problem Set 10.1 (page 624)

1. (a) Answers will vary; the *p* should extend into the lowest row and leave the upper 2 or 3 rows blank. One possibility is shown.

2. (a)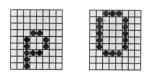

3. (a) ∠BAC will be a right angle if C is any of the circled points.

   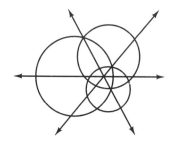

4. (a) Excluding zero angles, there are 10 (nondirected) angles:
   ∠APB, ∠APC, ∠APD, ∠APE, ∠BPC, ∠BPD, ∠BPE, ∠CPD, ∠CPE, ∠DPE.

5. $m(\angle AXB) = m(\angle AXE) - m(\angle BXE) = 180° - 132° = 48°$
   $m(\angle CXD) = m(\angle BXD) - m(\angle BXC) = 90° - 35° = 55°$
   $m(\angle DXE) = m(\angle BXE) - m(\angle BXD) = 132° - 90° = 42°$

7. The three lines are concurrent. A possible drawing is shown.

9. (a) Measurements will vary. For the figure shown in the textbook, $m(\angle A) \doteq 111°$, $m(\angle B) \doteq 88°$, $m(\angle C) \doteq 69°$, and $m(\angle D) \doteq 92°$. In general, opposite angles are supplementary:
   $m(\angle A) + m(\angle C) = 180°$ and $m(\angle B) + m(\angle D) = 180°$.

10. (a) Measurements will vary. For the figure shown in the textbook, $m(\angle AOB) \doteq 64°$, $m(\angle APB) \doteq 32°$, $m(\angle AQB) \doteq 32°$, and $m(\angle ARB) \doteq 32°$. In general, you should find $m(\angle APB) = m(\angle AQB) = m(\angle ARB)$ and this measure is half of the measure of ∠AOB.

11. (a) 58° 36' 45"
    $= 58° + \left(\frac{36}{60}\right)° + \left(\frac{45}{3600}\right)° = 58.6125°$

12. Ten times. A zero angle is formed at approximately 1:05, 2:11, 3:16, 4:22; 5:27, 6:33, 7:38, 8:44, 9:49, and 10:55.

13. (a) One complete turn, or 360°

(d) $\frac{1}{12}$ of a complete turn: $\frac{1}{12}(360°) = 30°$

14. (a) The minute hand is on the 12 and the hour hand is on the 3, so the hands form a right angle: 90°.

    (c) The minute hand is on the 6 and the hour hand is halfway between the 4 and the 5. The angle between two consecutive numbers is $\frac{1}{12}$ of a revolution, or 30°. So the angle is $(1.5)(30°) = 45°$.

16. Draw a horizontal ray $\vec{PQ}$ at P, in the opposite direction of $\vec{AB}$ and $\vec{CD}$. The opposite interior angles theorem then gives $m(\angle APQ) = 130°, m(\angle CPQ) = 140°$. Thus $m(\angle P) = 360° - 130° - 140° = 90°$.

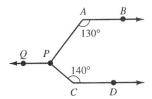

17. (a) The measures of the interior angles of a triangle add up to 180°. Therefore, $m(\angle 1) = 180° - 70° - 70° = 40°$.

    (c) The interior angles of a triangle add up to 180°, and a right angle has measure 90°, so $m(\angle 3) + 90° + 41° = 180°$. Therefore, $m(\angle 3) = 49°$.

18. (a) $x + x + 30° = 180°$ so $x = 75°$. The interior angles measure 75°, 75°, and 30°.

19. (a) No, because an obtuse angle has measure greater than 90° and adding two such measures would exceed 180° which is the sum of all three interior angle measures for any triangle.

20. (a) Zero intersection points if the five lines are parallel to each other.

21. (a) 6 lines

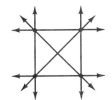

22. Notice that adjacent angles are not all composed from adjacent rays. But a pair of adjacent angles is determined by three rays and each collection of three rays determines a different pair of adjacent angles. So an equivalent question is: How many ways can a subset of 3 rays be chosen from 10 rays? $C(10, 3) = 120$ ways

24. The pencil turns through each interior angle of the triangle. Notice that each turn is in the same direction (counterclockwise). Since the pencil faces the opposite direction when it returns to the starting side, it has turned a total of 180°. This demonstrates that the sum of measures of the interior angles of a triangle is 180°.

25. (a) By trial and error, these are 10 taxi-segments. Notice that although there are infinitely many paths from A to B, the taxi-segments correspond to the paths that accomplish the journey in 5 blocks—that is, no backtracking. Any such journey must include 3 eastward blocks and 2 northward blocks. Since these may be undertaken in any order, the number of paths may also be determined as the number of ways to arrange the symbols E, E, E, N, N. This is $C(5, 2) = 10$.

26. The lights can always be turned to illuminate the whole plane. Here's one way to do this. Relabel the points if necessary so that A is at least as far south as B and B is at least as far south as C and D. That is, A is the southernmost point, and B is the next southernmost point. If A is west of B, point the light at A to the northeast, and point the light at B to the northwest. Otherwise, point the light at A to the northwest, and point the light at B to the northeast. This illuminates every point that is at least as far north as B. Similarly, point the remaining two lights southeast and southwest to illuminate the rest of the plane.

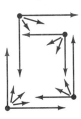

28. (a)

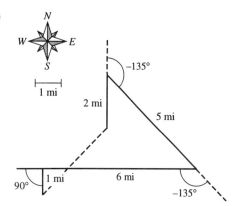

30. (a) A full revolution takes 24 hours. In 1 hour, the earth turns $\frac{1}{24}$ of a revolution, that is, $\frac{1}{24} \cdot (360°) = 15°$.

31. Extend the line $\overleftrightarrow{PS_2}$ downward to form a right triangle as shown.
Then $m(\angle 2) = 180° - 90° - 37° = 53°$, and $m(\angle 1) = 180° - 90° - 53° = 37°$. In general, the angle of latitude ($\angle 1$) has the same measure as the angle of elevation of Polaris—37°, in this case.

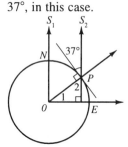

33. (a) Q, R, and S are collinear.

34. (a) The lines $\overleftrightarrow{BB'}$ and $\overleftrightarrow{DD'}$ intersect at a right angle at P.

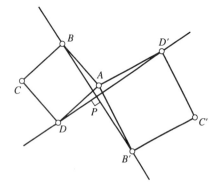

JUST FOR FUN Triangle Pick-Up Sticks (page 633)

Some answers may vary.

a. Remove 6 match sticks.

b. Remove 7 match sticks.

c. Remove 8 match sticks:

d. Remove 9 match sticks: This case is not possible, because it is not possible to form the desired regions using the 7 remaining match sticks. It is possible to form two triangles as shown, but the larger triangle does not form a triangular "region" because it actually contains two regions.

e. Remove 10 match sticks.

f. Remove 11 match sticks.

Problem Set 10.2 (page 644)

1.

	(a)	(b)	(c)	(d)	(e)	(f)	(g)	(h)	(i)	(j)	(k)	(l)
Simple curve	√		√	√	√	√	√	√	√			
Closed curve		√	√	√			√	√		√	√	√
Polygonal curve	√	√		√		√	√	√			√	√
Polygon		√				√	√			√		

2. (a) An example is shown. Note that the "sides" of this polygonal curve are the two vertical line segments and the two diagonal segments that connect the vertical segments. (If drawn in a different order, this could be regarded as a 6-sided polygonal curve.)

   (c) Any rectangle is acceptable. An example is

4. (a) Convex

   (b) Nonconvex, because there is a segment whose endpoints are in the figure that contains points not in the figure.

5. The shaded region is always convex.

   (a)

   (c)

6. (a) 6 regions: the interior of the circle, the exterior of the square, and the four corner regions.

   (c) 8 regions: the interior of the hexagon, the six triangular regions, and the exterior region.

7. The polygon has 6 sides, so the sum of the interior angle measures is
   $(6-2)(180°) = 720°$. Therefore,
   $2x + 5x + 5x + 5x + 5x + 2x = 720°$
   $24x = 720°$
   $x = 30°$.
   Since $2x = 60°$ and $5x = 150°$, the angles measure 60°, 150°, 150°, 150°, 150°, and 60°.

9. (a) The polygon has 5 sides, so the sum of the interior angle measures is
   $(5-2)(180°) = 540°$.

   (c) The polygon has 6 sides, so the sum of the interior angle measures is
   $(6-2)(180°) = 720°$.

10. Pictures will vary.

    (a) $(n-2)(180°) = 180°$
    $n - 2 = 1$
    $n = 3$
    The lattice polygon can be any triangle.

    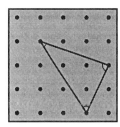

    (c) $(n-2)(180°) = 1440°$
    $n - 2 = 8$
    $n = 10$
    The lattice polygon can be any decagon.

    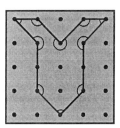

12. (a) 360°

(c) 0°

14. In each case, the measure of the interior angle is given by $\frac{(n-2)(180°)}{n}$, and the measures of the exterior angle and central angle are each given by $\frac{360°}{n}$ (or 180° minus the exterior measure).

	n	Interior Angle	Exterior Angle	Central Angle
(a)	7	$128\frac{4}{7}°$	$51\frac{3}{7}°$	$51\frac{3}{7}°$
(c)	5	108°	72°	72°

16. (a) $\frac{360°}{n} = 15°$ so $n = 24$.

18. (a) 8 regions. There are several ways to accomplish this. (See the answer to Problem 25 for additional explanation.)

19. (a) The right angle is located at A or B. Therefore, C could be any point (other than A or B) on either of the two lines drawn through A and through B which are perpendicular to $\overline{AB}$.

(b) The right angle is located at C. Therefore, by Example 10.2(b), C could be any point (other than A or B) on the circle with $\overline{AB}$ as diameter.

20. At a vertex, the interior angle and the conjugate angle add up to 360°. For an n-gon, the sum of all interior and all conjugate angles is $n \cdot 360°$. All the interior angles add up to $(n-2) \cdot 180°$, so all the conjugate angles add up to
$360°n - (n-2)180° = 360°n - 180°n + 360°$
$= 180°n + 360° = (n+2) \cdot 180°$.

22. Such a point S allows for n triangles to be formed, all with the vertex S. The sum of all the interior angles of these n triangles is $n \cdot 180°$ which is equal to the sum of all the interior angles of the n-gon plus 360° for the angles that surround the point S. Thus the sum of the interior angles of the n-gon is
$n \cdot 180° - 360° = (n-2) \cdot 180°$.

24. (a), (b)
For part (a), the number of pieces is one more than the number of points removed. For part (b), refer to the illustrations below.

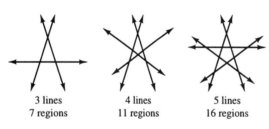

The partially completed table is shown.

	Number of Points or Lines Removed							
	0	1	2	3	4	5	6	7
Number of Pieces of Line	1	2	3	4	5	6	7	8
Number of Pieces of Plane	1	2	4	7	11	16		

(c) The pattern is to add the number of pieces of the line and the number of pieces of the plane for a value n to get the number of pieces of the plane with $n + 1$ lines. For example, $5 + 11$ gives 16. Then $6 + 16$ gives 22, and so on. Continuing in this fashion, 10 lines determine 56 pieces of the plane.

(d) Comparing to the sequence of triangular numbers 1, 3, 6, 10, 15, ..., we note that the number of regions is one more than the nth triangular number. That is, the number of regions is $1 + \dfrac{n(n+1)}{2}$.

25. Each new circle creates a new region each time it intersects a previously drawn circle. Since the new circle intersects each of the old circles in at most two points, this creates the following pattern:

Number of	
Circles	Regions
1	2
2	$2 + 2 \cdot 1 = 4$
3	$4 + 2 \cdot 2 = 8$
4	$8 + 2 \cdot 3 = 14$
5	$14 + 2 \cdot 4 = 22$
6	$22 + 2 \cdot 5 = 32$
7	$32 + 2 \cdot 6 = 44$
8	$44 + 2 \cdot 7 = 58$
9	$58 + 2 \cdot 8 = 74$
10	$74 + 2 \cdot 9 = 92$

27. $n = 4$ The vertices of a rhombus. Alternatively, the vertices of an equilateral triangle plus the center point

          • •   or •   •
        •   •   • •

$n = 5$ The vertices of a regular pentagon.

$n = 6$ The vertices of a regular pentagon plus the center point.

29. (a) The sum of the interior angles is 360°
$= m(\angle P) + m(\angle Q) + m(\angle R) + m(\angle S)$.
We also know $m(\angle P) = m(\angle R)$ and $m(\angle Q) = m(\angle S)$, so $360°$
$= m(\angle P) + m(\angle Q) + m(\angle P) + m(\angle Q)$
$= 2 \cdot (m(\angle P) + m(\angle Q))$, giving us $180°$
$= m(\angle P) + m(\angle Q)$.

(b) Draw $PQRS$ and extend $\overline{PQ}$ to form $\angle q$, as shown. $m(\angle Q) + m(\angle q) = 180°$ and $m(\angle P) + m(\angle Q) = 180°$ so $m(\angle q) = m(\angle P)$. $\angle q$ and $\angle P$ are corresponding angles, so segments $\overline{PS}$ and $\overline{QR}$ are parallel. $m(\angle q) = m(\angle P)$ and $m(\angle P) = m(\angle R)$ so $m(\angle q) = m(\angle R)$. $\angle q$ and $\angle R$ are alternate interior angles, so their congruence gives $\overline{PQ}$ parallel to $\overline{SR}$. Hence the figure is a parallelogram.

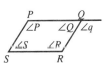

31. (a) Drawings will vary. Two possibilities are shown.

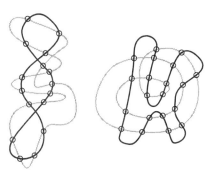

(b) The black curve determines various regions (before the red curve is drawn). These regions can be colored yellow and green in such a way that each yellow region is adjacent to only green regions and vice versa, where adjacent is defined as sharing a section of the boundary curve (not just isolated points). (Note: This assumes that the black curve intersects itself only in isolated points.) When the red curve is drawn, each crossing switches it from a yellow region to a green region or vice versa. Since the red curve is closed, it must eventually get back to the same colored region where it started—so it crosses the black curve an even number of times.

33. The square could fall through the hole. This is because a square's side length is less than the length of its diagonal. This cannot happen for a circle, because the hole's opening is slightly smaller than the diameter of the cover.

35. (a) Drawings will vary. An example is given in the textbook. *SQRE* is a square.

    (b) No. As in the example shown below, *SQRE* will still be a square.

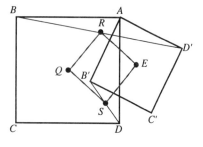

37. Note that (a) through (h) can be accomplished by using the procedure
    to polygon :n sides :side
    with appropriate choices of the values of the two variables. This procedure is given in Example C.9 on page 992. Alternately, the following procedures can be used. The side lengths are chosen arbitrarily. It may need to be decreased for some many-sided shapes so that the result will fit on the screen.

    (a) ```
to pentagon
    repeat 5 [fd 20 rt 72]
end
```

    (c) ```
to octagon
    repeat 8 [ fd 20 rt 45 ]
end
```

    (e) ```
to twenty
    repeat 20 [fd 5 rt 18]
end
```

    (g) ```
to sixty
    repeat 60 [ fd 2 rt 6 ]
end
```

    (i) A regular 360-gon will be a good approximation of a circle. See part (h).

38. (a) Answers will vary. If we travel clockwise, then at the interior reflex angles we need to turn left $180° - 144° = 36°$, and at the "points" we need to turn right $180° - 72° = 108°$. This may be accomplished by the following procedure.

    ```
to openstar.5
    repeat 5 [ fd 30 lt 36 fd 30 rt 108]
end
```

40. If the angles are $36x$, $36y$, and $36z$, then $36x + 36y + 36z = 180°$. This gives $x + y + z = 5$, so the three numbers are 1, 2, and 2 *or* 1, 1, and 3. This gives rise to the two triangles shown.

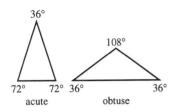

42. One angle is 90°; call the other two angles ∠A and ∠B. The interior angles of a triangle add up to 180°, so $90° + m(\angle A) + m(\angle B) = 180°$, or $m(\angle A) + m(\angle B) = 90°$. Thus A and B are complementary.

JUST FOR FUN Dissection Delights (page 657)

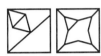

Problem Set 10.3 (page 663)

1. Many different tilings can be formed. For example:

   (a) The pattern shown consists of squares within squares—all formed from identical triangles.

   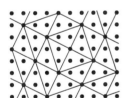

2. (a) As shown in Example 10.14, any quadrilateral can tile the plane.

3. (a) Squares, pentagons, hexagons, heptagons, octagons

   (b) Actually, it is a "real" tiling by "fake" regular polygons. Many vertex figures that appear in the tiling cannot correspond to regular polygons. For example, a pentagon, hexagon, and octagon meet at each of many vertices. If these were regular figures, they would have interior angles of 108°, 120°, and 135°, respectively. But $108° + 120° + 135° = 363° \neq 360°$, so these cannot be all regular polygons.

5. The interior angle of a square is 90° and for a regular pentagon is $\frac{(5-2)(180°)}{5} = 108°$ and for a regular 20-gon is $\frac{(20-2)(180°)}{20} = 162°$.
Then,
$90° + 108° + 162° = 360°$.

7. Various tilings can be formed.

8. Each pentomino tiles the plane.

9. (a)

   (b) The five tetrominoes cannot tile a 4 by 5 rectangle. To see why, color the rectangle in a checkerboard pattern of 10 red and 10 black squares. All but the T-shaped tetromino cover 2 red and 2 black squares. The T-tetromino covers 3 of one color and 1 of the other color. Thus, the five tetrominoes together would cover 11 red squares and 9 black squares (or vice versa).

11. The hexagon will tile using translations only. No rotation is necessary.

13. Only the semiregular tiling in which each vertex is surrounded by a hexagon and four equilateral triangles will appear to be different. To see why, shade the six large equilateral triangles, each made up of four small equilateral triangles, about any hexagon. The "saw blade" formed has an orientation that is clearly reversed in a mirror image. Furthermore, this orientation is the same for every hexagon in the tiling.

It may be tempting to make a similar argument for the tiling in which each vertex is surrounded by two squares and three triangles.
A "saw blade" like the one above is formed by considering the rhombuses (each composed of two triangles) that surround a square. The argument fails in this case because the saw blades surrounding the
purple squares have the opposite orientation of the saw blades surrounding the green squares, so that (when we ignore colors) the overall pattern is the same as its reflection.

17. Here's one way to make a tiling 12-gon, by modifying a square tile:

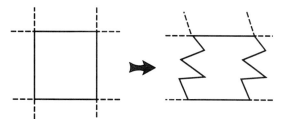

Similarly, modifying a pentagon with opposite parallel congruent sides will form a 13-gon that tiles:

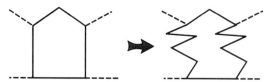

The same idea can be used for any $n \geq 6$, using a modified square for even n, and a modified pentagon for odd n.

18. Some tilings will vary.

(a) Any four identical triangles can be arranged to form a new triangle that has the same shape, but is twice as large.

140 *Chapter 10:* Mathematical Reasoning for Elementary Teachers (c)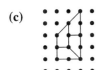

20. (a) Four Sphinxes can be arranged to create a larger sphinx.

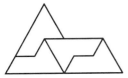

21. (a) A second generation Chevron is shown that uses 25 original Chevrons. It is even easier to construct a second generation that uses 36 of the original Chevrons.

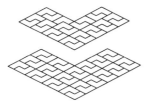

(b) These points could be connected to form a square grid. (Note that A, C, E, and G are all translates of each other.)

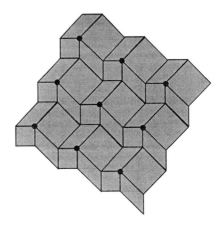

23. (a) The translates of V could be connected to form a square grid.

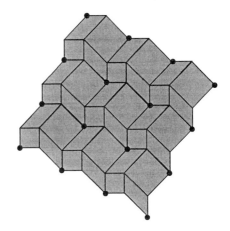

24. (a) Answers will vary. The procedure to **sq.tile**, given below, begins by picking up the pen, moving it to the bottom right corner of the screen, and then putting the pen down. The bottom row of squares is traced from right to left, and then the next row is traced from left to right. Then four more rows are created, producing the result shown below.

```
to square :side
  repeat 4 [fd :side rt 90 ]
end

to sq.tile :side
  pu bk 100 lt 90 bk 100 pd
  repeat 3 [ repeat 10 [ square:side fd:side ]
      rt 90 fd 2 *:side rt 90
      repeat 10 [ square:side fd:side ]
      lt 180 ]
end
```

Chapter 10: Calvin T. Long and Duane W. DeTemple **141**

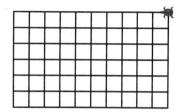

25. (a) Answers will vary. The procedure **tile.pos**, given below, begins by picking up the pen and moving it to the upper left corner. The top row of parallelograms is created from left to right, and then the next row is created from right to left. This is then repeated.

```
to tile.par
  pu fd 50 lt 90 fd 100 rt 180 pd
  repeat 2 [ repeat 12 [ parallelogram fd 30]
      rt 110 fd 80 rt 70
          repeat 12 [ parallelogram fd 30 ] rt 180 ]
end
```

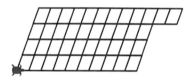

28. Letters above the line are formed with straight pen strokes while letters below the line involve curved pen strokes.

29. Let A, B, and C be the vertices of an equilateral triangle, and D the center (centroid) of the triangle. (These four points can be labeled in any order.)

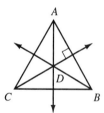

Problem Set 10.4 (page 674)

1. (a) Yes, it is traversable because it has no odd vertices. *AHGFEDCBHFDBA* is one Euler path.

(c) No, it is not traversable because it has more than two odd vertices (A, B, C, D, E, and F).

(e) Yes, it is traversable because it has only two odd vertices. The odd vertices, C and D, must be the endpoints of any Euler path. One Euler path is *CEAEFBFDBACD*, as shown below. (Note that there are two distinct edges joining A and E, and two distinct edges joining B and F.)

2. (a)

Figure	D = total degree	E = number of edges
(a)	2 + 4 + 2 + 4 + 2 + 4 + 2 + 4 = 24	12
(b)	2 + 3 + 4 + 2 + 2 + 3 + 4 + 2 = 22	11
(c)	3 + 3 + 3 + 3 + 3 + 3 + 6 = 24	12
(d)	2 + 4 + 2 + 4 + 2 + 4 + 4 + 4 + 4 = 30	15
(e)	4 + 4 + 3 + 3 + 4 + 4 = 22	11
(f)	4 + 3 + 4 + 5 + 2 + 5 + 4 + 3 = 30	15

3. (a) Yes, since exactly two vertices are odd and the network is connected.

   (b) $D = 2 + 2 + 4 + 8 + 3 + 6 + 6 + 1 = 32$. Since $D = 2E$, there are 16 edges.

5. (a) Draw a vertex for each land mass and an edge for each bridge.

   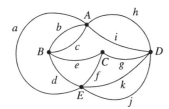

6. (a)

   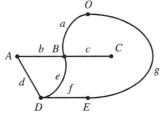

7. (a) There are 6 vertices, 7 regions (including the exterior region), and 11 edges. Therefore, $V = 6$, $R = 7$, and $E = 11$.

9. (a) No. Vertex 1 is of degree 5 but no vertex in the other network has degree 5.

   (c) Yes. Since B and 4 are the only vertices of degree 4, $B \leftrightarrow 4$. The vertices of degree 2 are A, E, 2, and 3, so these must correspond somehow. We try $A \leftrightarrow 3$ and $E \leftrightarrow 2$. Now, notice that A is connected to B and C, and its corresponding vertex 3 is connected to 4 and 5. We already know that $B \leftrightarrow 4$, so we conclude that $C \leftrightarrow 5$. The two remaining vertices must correspond, so $D \leftrightarrow 1$. After verifying that the edges correspond correctly, we can conclude that the networks are isomorphic. The one-to-one correspondence is $A \leftrightarrow 3$, $B \leftrightarrow 4$, $C \leftrightarrow 5$, $D \leftrightarrow 1$, $E \leftrightarrow 2$.

   Note: Another appropriate one-to-one correspondence exists. It is $A \leftrightarrow 2$, $B \leftrightarrow 4$, $C \leftrightarrow 1$, $D \leftrightarrow 5$, $E \leftrightarrow 3$.

10. (a) The two paths can be combined to give a new path that goes from one of the two vertices to the other and back again. This would be a closed path with distinct edges.

11. Since all trees are connected, a traversable tree is one that has zero or two odd vertices. But every tree with two or more vertices has at least two vertices of degree 1. So a nontrivial traversable tree must have exactly two vertices of degree 1 (and no other odd vertices). The only way this is possible is if the tree consists of a number of vertices connected in a row, like the network shown, or a single vertex with no edges.

13. Consider a network with a vertex for each person and an edge corresponding to each handshake. The number of edges joining any two people would be the number of times they shook hands. By the result stated in Problem 12, the number of odd vertices in this network is even. Therefore, there are an even number of people who have shaken hands an odd number of times.

15. It is always possible. Requiring that each edge is traced exactly twice is equivalent to duplicating each edge in the network and asking if the new "double edge" network is traversable. Since doubling the edges at each vertex always creates an even vertex, the "double edge" network has no odd vertices. Since the network is also connected, this means it is traversable and hence the original network can be traced as required.

17. (a) Drawings will vary. There will be 14 regions.

    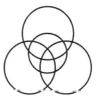

18. (a) ••••• •–< +

19. Answers will vary. In the original configuration, current can flow from any pin to any other pin. This just means that the network is connected. Thus, any 7 of the original 12 wires that produce a connected network will produce the desired result. An example is HA, AB, BG, GF, FD, DC, DE.

    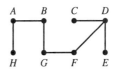

21. Drawings will vary. A square always results.

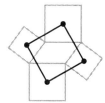

23. The supplement of ∠x measures 150° (by corresponding angles along a transversal of parallel lines), so x = 180° − 150° = 30°. The angles in the upper triangle are x, y, and 180° − 75° = 105°, so y = 180° − x − 105° = 180° − 30° − 105° = 45°. Since ∠z corresponds to the supplement of ∠y along a transversal of parallel lines, we can conclude that z = 180° − y = 180° − 45° = 135°. In summary, x = 30°, y = 45°, and z = 135°.

JUST FOR FUN   Space Out for Success! (page 682)

Form a tetrahedron with the 6 matches.

Problem Set 10.5 (page 689)

1. (a) Polyhedron. It is a simple closed surface formed from planar polygonal regions.

   (c) Polyhedron. Note that a polyhedron need not be convex.

   (e) Not a polyhedron. It is not a simple closed surface because it has two interior regions of space.

2. (a) Pentagonal prism (and possibly a right pentagonal prism)

   (c) Oblique circular cone

   (e) Right rectangular prism (assuming all angles are right angles)

3. (a) 4. There are 4 faces, and each is in a different plane.

   (c) A, B, C, D

5. (a) Note that the tetrahedron includes the lower back left corner.

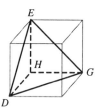

6. This problem may be solved by mental visualization or by actually constructing a model.

   (a)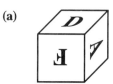

7. (a) 45° since the dihedral angle between the adjacent sides of the cube is 90° and another pyramid would fit into the gap between the original pyramid and a vertical side of the cube. This can also be determined by visualizing a cross section half-way into the cube and parallel to the cube's front face. The pyramid forms a right isosceles triangle in this cross section.

   (b) Filling the cube with six such pyramids, one sees that three pyramids sit around an edge formed by the lateral sides of a pyramid. Thus three copies of the dihedral angle give a full revolution of 360° around this edge, so the dihedral angle measures 120°.

9. (a) A triangle. The exact shape of the triangle depends on how the plane is oriented.

10. Each description below refers to the cube shown.

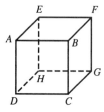

   (a) Possible. For example, the plane that contains A, C, and F gives a cross section that is the equilateral triangle △ACF.

(c) Possible. For example, the plane that contains A, D, and F gives a cross section that is the rectangle $ADGF$.

(e) Possible. For example, the plane that contains the midpoints of $\overline{AE}$, $\overline{EF}$, and $\overline{DC}$ forms a regular hexagon whose vertices are the midpoints of $\overline{AE}$, $\overline{EF}$, $\overline{FG}$, $\overline{GC}$, $\overline{CD}$, and $\overline{DA}$.

11. (a) A cube is a prism.

(b) A tetrahedron is a pyramid.

13. (a) Pentagonal double pyramid: $V = 7$, $F = 10$, $E = 15$, $V + F = 17 = E + 2$, so Euler's formula holds.
Hexagonal antiprism: $V = 12$, $F = 14$, $E = 24$
$V + F = 26 = E + 2$, so Euler's formula holds.

14. (a) Create the net by "unfolding" the figure.

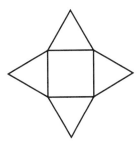

15. (a) $V = 10$, $F = 12$, $E = 20$
$V + F = 22 = E + 2$, so Euler's formula holds.

16. It is not possible to construct this figure using line segments. Consider the plane P through A, B, and C. Since C is in P and $\overline{CD}$ passes behind $\overline{AB}$ (which is also in P), D must be behind P. Then the segment DE starts behind P at D, passes in front of P as it crosses $\overline{AB}$, and then passes behind P as it crosses $\overline{BC}$. Since no line can cross back and forth through a plane, the space "pentagon" figure shown is an impossible figure. (Of course, it can be created if we allow the segments to become curves.)

17. (a) Many examples are possible. The figure shown could represent the outline of a 3-by-6 rectangular card with a 90° fold at its center.

19. (a) Suppose the faces of a polyhedron consist of a p-gon, a q-gon, an r-gon, and so on. Since each edge of the polyhedron borders two faces, the sum $p + q + r + \cdots$ is twice the number of edges. That is, $p + q + r + \cdots = 2E$. Since there are F faces and $p, q, r, \ldots$ are all 3 or greater, we get $2E \geq 3 + 3 + \cdots = 3F$.

(b) Each of the V vertices of a polyhedron is the endpoint of 3 or more edges that meet at the vertex. Thus, $3V$ is less than or equal to the total number of ends of the edges. But each of the E edges has 2 ends, so there are $2E$ ends of edges. We see that $3V \leq 2E$.

(c) Adding $3V \leq 2E$ and $3F \leq 2E$ shows that $3V + 3F \leq 4E$. But $V + F = E + 2$ (Euler's formula), so $3V + 3F = 3E + 6$. Comparing this to the inequality, we see that $3E + 6 \leq 4E$. Subtracting $3E$ from both sides shows that $6 \leq E$.

(d) Suppose $E = 7$. Since $3F \leq 2E = 14$, we see that F is no larger than 4 ($F \geq 5$ would give $3F \geq 15$). Similarly, $3V \leq 2E = 14$ means that $V \leq 4$. Since both $V \leq 4$ and $F \leq 4$, then $V + F \leq 8$. But $V + F = E + 2$ (Euler), and $E = 7$, so $V + F = 9$. This contradicts $V + F \leq 8$, so our assumption $E = 7$ is not possible.

(e) A pyramid with a base of 3, 4, 5, ..., n, ... sides has 6, 8, 10, ..., $2n$, ... edges, respectively. Slicing off a tiny corner at one vertex somewhere on the base of the pyramid adds three new edges, giving us polyhedra with 9, 11, 13, ..., $2n + 3$, ... edges. All together, the pyramids and pyramids with a truncated base corner give us polyhedra with 6, 8, 9, 10, 11, ... edges.

21. (a)

Tetrahedron

(b)

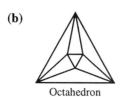
Octahedron

(c)

Icosahedron

23. **(a)** The pyramid must have a base and five other sides, so the base is a pentagon.

(b) The double pyramid must have three faces above and three faces below the center polygon, so the center polygon is a triangle.

(c) We show Schlegel diagrams (see Problem 21) for all seven types. The pyramid in part (a), viewed from above, would look like the left hand figure below. The double pyramid in part (b) is shown on the right. The second figure shown represents the "cube" type, with all quadrilateral faces. We may verify that all seven figures represent different types by counting how many of each kind of face are in each figure.

5 triangles   6 quadrilaterals   2 triangles   2 triangles   4 triangles   3 triangles   6 triangles
1 pentagon   4 quadrilaterals   2 quadrilaterals   1 quadrilateral   2 quadrilaterals
                                                     2 pentagons   1 pentagon   1 pentagon

25. **(a)** When folded, points A, C, E, and G all correspond to the apex of the pyramid. Thus, for example, $\overline{AB}$ and $\overline{BC}$ represent the same edge from the pyramid and so $AB = BC$. Likewise, $CD = DE$, $EF = FG$, and $GH = HA$.

To see why $\overline{AP}$ is perpendicular to $\overline{BH}$, let R be the plane that is perpendicular to $\overline{BH}$ and contains A, and consider what happens to point A as $\triangle ABH$ is folded upward to form the pyramid. As the figure is folded along $\overline{BH}$, A moves along a circle in plane R.

This circle contains the apex of the pyramid and it also contains the point that would be the apex if the pyramid were turned upside down. Since P is on the line segment that connects these points (indeed, it is the midpoint), we conclude that P is in plane R. Therefore, in the net, $\overline{AP}$ is in plane R. Since R is perpendicular to $\overline{BH}$, we conclude that $\overline{AP}$ is perpendicular to $\overline{BH}$. Likewise, $\overline{CP}$, $\overline{EP}$, and $\overline{GP}$ are perpendicular to $\overline{BD}$, $\overline{DF}$, and $\overline{FH}$, respectively.

26. The axis of all the hinges on a door must be along the intersection of the planes of the wall and the plane of the opened door. Since planes meet in a line, the axes of the hinges must be along a single line.

29. Since $\angle x$ and a 130° angle are alternate interior angles along a transversal of parallel lines, $m(\angle x) = 180° - 130° = 50°$. Then, by looking at the large triangle, $m(\angle y) = 180° - 80° - 50° = 50°$. Finally, since $\angle y$ corresponds to the supplement of $\angle z$ along a transversal of parallel lines, $m(\angle z) = 180° - 50° = 130°$. In summary, $m(\angle x) = 50°$, $m(\angle y) = 50°$, and $m(\angle z) = 130°$.

30. (a) Draw a convex pentagon, hexagon, heptagon, and octagon to see how many diagonals each has.

n	3	4	5	6	7	8
diagonals	0	2	5	9	14	20

    (b) The number of diagonals in a convex n-gon is $t_{n-2} - 1 = \frac{(n-2)(n-1)}{2} - 1 = \frac{n(n-3)}{2}$. This is expected because each vertex is an endpoint for $n - 3$ diagonals, so the total number of endpoints of diagonals is $n(n-3)$. Since each diagonal has two endpoints, there are $\frac{n(n-3)}{2}$ diagonals.

    (c) A dodecagon has 12 vertices, each of which is an endpoint for 9 diagonals. Thus there are $12 \cdot 9 = 108$ endpoints of diagonals, or 54 diagonals.

    (d) $\frac{100 \cdot 97}{2} = 4850$ diagonals

31. (a) Answers will vary. One possibility is shown.

CLASSIC CONUNDRUM
An Unexpected Bisector
(page 694)

By adding the three right triangles shown, we see that $\overline{PQ}$ is part of the diagonal of a square, and therefore it bisects $\angle P$.

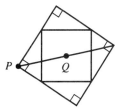

Chapter 10 Review Exercises (page 696)

1. (a) $\overleftrightarrow{AC}$

   (b) $\overline{BD}$

   (c) AD

   (d) $\angle ABC$ or $\angle CBA$ (or simply $\angle B$). Note that all other angles shown, when considered as a union of two rays, contain D.

   (e) $m(\angle BCD)$, $m(\angle DCB)$, or $m(\angle C)$

   (f) $\overrightarrow{DC}$

2. (a) $\angle BAD$

   (b) $\angle BCD$

   (c) $\angle ABC$, $\angle ADC$

3. (a) $180° - 37° = 143°$

   (b) $90° - 37° = 53°$

4. Since $\angle p$ and a 125° angle are adjacent supplementary angles, $p = 180° - 125° = 55°$. Since l and m are parallel, the small triangle is a right triangle, and so $q = 180° - 90° - p = 35°$. Since r corresponds to p along a transversal of parallel lines, $r = p = 55°$. Since $\angle r$ and $\angle s$ are adjacent supplementary angles, $s = 180° - 55° = 125°$. In summary, $p = 55°$, $q = 35°$, $r = 55°$, and $s = 125°$.

5. Since ∠x and a 135° angle are adjacent supplementary angles, x = 180° − 135° = 45°. Since the sum of the measures of the angles in a triangle is 180°, y = 180° − 102° − x = 180° − 102° − 45° = 33°. Since ∠y and ∠z are adjacent supplementary angles, z = 180° − y = 180° − 33° = 147°.
In summary, x = 45°, y = 33°, and z = 147°.

6. (a) iv

   (b) i

   (c) vi

   (d) v

   (e) ii

   (f) iii

7. (a) No, because obtuse angles have measure greater than 90° and the sum of the three interior angles of a triangle is 180°. Two obtuse angles would produce a sum greater than 180°.

   (b) Yes, try angles of 100°, 100°, 100°, and 60°.

   (c) No, because acute angles have measure less than 90°, so if all the interior angles are acute the sum would be less than 4 · 90° = 360°. But the sum of the interior angles of a quadrilateral must be 360°.

8. The figure is a hexagon, so the interior angles add up to (6 − 2)(180°) = 720°. Therefore, 3x + 3x + 3x + x + 5x + x = 720°, which means 16x = 720°, or x = 45°. The angles are 135°, 135°, 135°, 45°, 225°, and 45°.

9. Us the Total Turtle Trip Theorem. He turns through 360°.

10. The angles are 60° for the triangle, 90° for the square, and 120° for the hexagon. The angles of the four polygons must add up to 360°, so the fourth angle is 90°, and therefore the fourth polygon is a square.

11. The two possibilities are that the squares are consecutive, or they are not.

12. Use the 180° rotations of the tile about the midpoints of the sides.

    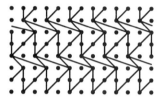

13. (a) It has four odd vertices (A, B, C, and E).

    (b) An edge should be added between any two of the vertices A, B, C, and E. Then the Euler path must begin and end at the remaining two of these vertices. If an edge is added that joins A and C, then one Euler path is BADACFCBFEDBE, as shown below. (Note that these are two edges joining A and D, and two edges joining C and F.) Many other Euler paths are possible.

    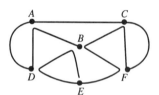

14. Construct a network with vertices A, B, C, D, and E and edges corresponding to bridges. Since just two vertices, A and D, have odd degree, there is an Euler path. The Euler path corresponds to a walking path which crosses each bridge exactly once.

    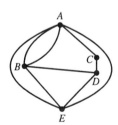

15. There are 11 vertices, 7 regions (including the exterior region), and 16 edges. Therefore, $V = 11$, $R = 7$, and $E = 16$. Then $V + R = 18$ and $E + 2 = 18$. Since these quantities are equal, it is true that
$V + R = E + 2$, so Euler's formula holds.

16. (a) 6, one for each face of the cube.

    (b) $\overline{CD}, \overline{EF}, \overline{GH}$

    (c) $\overline{CG}, \overline{DH}, \overline{EH}, \overline{FG}$

    (d) 45°, the same as $m(\angle DAH)$, since $\overrightarrow{AD}$ and $\overrightarrow{AH}$ each lie in one of the two planes and are perpendicular to $\overleftrightarrow{AB}$, the line of intersection of the two planes.

17. Square right prism; triangular pyramid (or tetrahedron); oblique circular cylinder; sphere; hexagonal right prism

18. (a)

    (b)

    (c)

19. (a) See Figure 10.44 on page 683.

    (b) There are 6 vertices, 8 faces, and 12 edges, so $V = 6$, $F = 8$, and $E = 12$. Thus $V + F = 14$ and $E + 2 = 14$. Since these quantities are equal, Euler's formula holds.

20. Note that $F = 14$ and $E = 24$. Then Euler's formula, $V + F = E + 2$, means that $V + 14 = 24 + 2$, so $V = 12$. These are 12 vertices.

Chapter 10 Test (page 699)

1.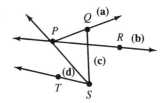

2. (a) A, C

   (b) F

   (c) D

   (d) E

   (e) B

3. Many examples of each are possible.

   (a)

   (b)

   (c)

   (d)

4. (a) True. The sides of a square are all the same length.

   (b) False. The sum of the measures of a right angle, an obtuse angle, and a third angle would be greater than 180°.

   (c) True. Any triangle with at least two congruent sides is isosceles.

   (d) True. The sides of a square can be separated into two pairs of congruent adjacent sides.

5. (a) Decagon, since $\frac{360°}{36°} = 10$.

   (b) Octagon, since $\frac{360°}{45°} = 8$.

   (c) Nonagon, since the exterior angles measures $180° - 140° = 40°$, and $\frac{360°}{40°} = 9$.

6. The small triangle on the right has angles 40°, 180° − 140° = 40°, and 180° − r, so 40° + 40° + (180° − r) = 180°, which gives r = 80°. Since ∠s corresponds to the supplement of ∠r along a transversal of parallel lines, s = 180° − r = 180° − 80° = 100°. Since ∠t corresponds to the supplement of a 140° angle along a transversal of parallel lines, t = 180° − 140° = 40°. In summary, r = 80°, s = 100° and t = 40°.

7. (a) Yes. Any triangle will tile the plane.

   (b) Yes. Any quadrilateral will tile the plane.

   (c) Yes. Any quadrilateral will tile the plane.

   (d) Yes. Any pentagon with two pairs of parallel sides will tile the plane.

   (e) Yes. Any hexagon with three pairs (or even one pair) of opposite parallel sides of the same length will tile the plane.

   (f) No. The plane cannot be tiled with any convex polygon having seven or more sides.

8. A vertex figure must use two octagons and a square, since 135° + 135° + 90° is the only combination that uses both 135° and 90° angles and adds up to 360°.

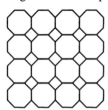

9. The total turn angle is 1080°, so the turn at each of the eight vertices is $\frac{1080°}{8} = 135°$. This means that the interior angle at each point is 180° − 135° = 45°.

10. The average interior angle measure for an n-gon is $\frac{(n-2)(180°)}{n}$. We want this value to be 174°, so $\frac{180(n-2)}{n} = 174$. This gives 180n − 360 = 174n, which implies n = 60. The polygon has 60 sides.

11. Note that (i) has 2 odd vertices, (ii) has 2 odd vertices, (iii) has 4 odd vertices, and (iv) has no odd vertices.

    (a) (iii), as this is the only network with 4 or more odd vertices

    (b) (iv), as this is the only network with no odd vertices

    (c) (i), (ii), and (iv) are traversable, since these networks have zero or two odd vertices.

    (d) (iii) requires an additional edge for traversability. The edge can join any two of the four odd vertices (that is, the three vertices on the inner circle and the vertex at the top of the diagram).

12. (a) Since E = 11 and R = 7, Euler's formula gives V + R = E + 2, or V + 7 = 11 + 2. Therefore, V = 6. There are 6 vertices.

    (b) Many such networks can be drawn. One possibility is shown.

13. The same as an interior angle of a regular pentagon, which is $\frac{3 \cdot 180°}{5} = 108°$.

14. (a) 24. Each edge borders a square face and a triangular face. So just count the edges of all the square faces (that is, 6 · 4 = 24) or the edges of all the triangular faces (that is, 8 · 3 = 24).

    (b) 12. Either count in the diagram (4 on the top, 4 on the bottom, and 4 in between) or use Euler's formula, V + F = E + 2.

15. (a) Diagrams will vary.

    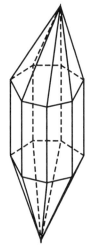

    (b) V = 16, F = 21, E = 35

    (c) V + F = 16 + 21 = 37, and E + 2 = 35 + 2 = 37. Since these quantities are equal, Euler's formula holds.

Chapter 11

JUST FOR FUN   **Twice Around a Triangle (page 715)**

P'' coincides with the original point P.

Problem Set 11.1 (page 717)

1. (a) $L \leftrightarrow K$, $H \leftrightarrow W$, $S \leftrightarrow T$

   (b) $\overline{LH} \leftrightarrow \overline{KW}$, $\overline{HS} \leftrightarrow \overline{WT}$, $\overline{SL} \leftrightarrow \overline{KT}$

   (c) $\angle L \leftrightarrow \angle K$, $\angle H \leftrightarrow \angle W$, $\angle S \leftrightarrow \angle T$

   (d) $\triangle LHS \cong \triangle KWT$

3. (a) Draw a line segment and mark a point E. Set the compass to AB and determine a point G on the line segment with an arc centered at E. Set the compass to CD and draw an arc centered at G, away from E. This determines a point F on the line segment. Thus, $EF = x + y$.

   (b) Draw a line segment and mark a point E. Set the compass to AB and determine a point G on the line segment with an arc centered at E. Set the compass to CD and draw an arc centered at G, towards E. This determines a point F on the line segment. Thus $EF = x - y$.

5. (a) Use the ruler to draw segment $\overline{AB}$ of length 5 cm. Use the protractor to construct $\angle A$ with measure 28°. Along the terminal side, use the ruler to locate a point C that is 5 cm away from A. Draw segment $\overline{BC}$ to complete the triangle.

   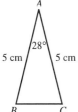

   There is only one possible triangle.

   (c) This is impossible by the triangle inequality since $2 + 5 < 8$.

   (e) Use the protractor to draw a right angle at C. Use the ruler to draw segment $\overline{AC}$ of length 6 cm along one side of the angle and then draw segment $\overline{BC}$ of length 4 cm along the other side. Draw segment $\overline{AB}$ to complete the triangle.

   There is only one possible triangle.

7. (c) $\overline{AB} \cong \overline{EF}$, $\overline{BC} \cong \overline{FD}$, and $\overline{AC} \cong \overline{ED}$. $\triangle ABC \cong \triangle EFD$ by the SSS property.

   (e) $\angle BAD \cong \angle CAD$, $\angle BDA \cong \angle CDA$, and $\overline{AD} \cong \overline{AD}$. $\triangle ABD \cong \triangle ACD$ by the ASA property.

   (g) $\angle DAC \cong \angle BCA$, $\overline{AC} \cong \overline{CA}$, and $\overline{CD} \cong \overline{AB}$, but no conclusion is possible since there is no "SSA" property.

   (i) $\overline{AB} \cong \overline{DE}$, $\overline{AC} \cong \overline{DC}$, and $\angle ACB \cong \angle DCE$, but no conclusion is possible since there is no "SSA" property.

8. Let $\triangle ABC$ be equilateral. Since $AB = BC$, it follows from the isosceles triangle theorem that $\angle A \cong \angle C$. In the same way, since $BC = CA$ it follows that $\angle B \cong \angle A$. Thus all three angles are congruent.

10. Let AB be the length of the hypotenuse. Construct the equilateral triangle shown. Since $\triangle ABC$ is equilateral, $AB = BC = AC$. $\overline{AB} \cong \overline{CB}$ and $\overline{BM} \cong \overline{BM}$, so $\triangle ABM \cong \triangle CBM$ by the HL theorem. Since $\triangle ABM \cong \triangle CBM$, $\overline{AM} \cong \overline{CM}$ or $AM = CM$. Thus, $2AM = AC = AB$, so $AM = \frac{1}{2} AB$.

    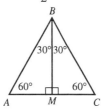

11. (a) Since alternate interior angles between parallel lines are congruent, $\angle ABD \cong \angle CDB$ and $\angle ADB \cong \angle CBD$. $\overline{DB} \cong \overline{BD}$, thus, $\triangle ABD \cong \triangle CDB$ by the ASA property.

13. (a) Refer to the diagram in problem 12. By problem 12, $AM = CM$ and $MB = MD$. $\angle AMB \cong \angle CMB \cong \angle CMD \cong \angle AMD$ since they are right angles, so $\triangle AMB \cong \triangle CMB \cong \triangle CMD \cong \triangle AMD$ by the SAS property. Thus, $\overline{AB} \cong \overline{CB} \cong \overline{CD} \cong \overline{AD}$ and $ABCD$ is a rhombus by definition.

14. Let a be the measure of the third side in centimeters. By the triangle inequality, $a < 4 + 9$ or $a < 13$. Also by the triangle inequality, $9 < a + 4$, so $5 < a$. Thus, the third side is longer than 5 cm and shorter than 13 cm.

15. (a) The sum of the three sides is 14 cm, so the fourth side must be less than 14 cm: $0 < s < 14$ cm, where s is the length of the fourth side.

17. Place a corner of the rectangular sheet of paper at a point C on the circle and mark the points A and B where the edges of the paper cross the circle. By Thale's theorem, $\overline{AB}$ is a diameter of the circle. Repeating the procedure at a second point C' will allow you to construct a second diameter $\overline{A'B'}$. The center of the circle is where the two diameters intersect.

19. (a) Since $AB = AE$, $\triangle ABE$ is isosceles. Therefore, $\angle B \cong \angle E$ by the isosceles triangle theorem.

   (b) Since $AC = AD$, $\triangle ACD$ is isosceles. Therefore, $\angle ACD \cong \angle ADC$ by the isosceles triangle theorem.

   (c) $\angle ACB$ and $\angle ACD$ are supplementary, as are $\angle ADC$ and $\angle ADE$. Since $\angle ACD \cong \angle ADC$,
   $m(\angle ACB) = 180° - m(\angle ACD)$
   $= 180° - m(\angle ADC) = m(\angle ADE)$, so $\angle ACB \cong \angle ADE$. Using $\angle B \cong \angle E$ from part (a) and $AC = AD$, $\triangle ABC \cong \triangle AED$ by the AAS property.

   (d) Using the result of part (c), $\overline{BC} \cong \overline{ED}$ because corresponding parts of congruent triangles are congruent. Thus $BC = DE$.

21. (a) Using $\angle A \cong \angle A$ and the given conditions, $\triangle ABC \cong \triangle ADE$ by the ASA property. The conditions are sufficient.

   (c) Using $\angle A \cong \angle A$ and the given conditions, $\triangle ABD \cong \triangle ADE$ by the SAS property.

22. Since $\triangle ABC$ is equilateral, the triangle is equiangular by problem 12, so $\angle A \cong \angle B \cong \angle C$. $AB = BC = CA$ by the definition of an equilateral triangle. Since $AD = BE = CF$ is given, $AB - AD = BC - BE = CA - CF$, or $BD = CE = AF$. $\triangle ADF \cong \triangle BED \cong \triangle CFE$ by the SAS property. Thus, $\overline{FD} \cong \overline{DE} \cong \overline{EF}$, so $\triangle DEF$ is equilateral.

24. If $AB = CD = a$, $BC = AD = b$, and $AC = BD = c$, then each face of the tetrahedron is a triangle with sides of length a, b, and c. By the SSS property, the faces of the tetrahedron are congruent to one another.

26. By the triangle inequality, $QT < PQ + PT$ and $SR < ST + TR$. Note that $QT = QS + ST$ and $PR = PT + TR$. Combining inequalities, $QT + SR < PQ + PT + ST + TR$, or $QS + ST + SR < PQ + ST + PR$, so $QS + SR < PQ + PR$.

29. (a) The angles at the vertices of a quadrilateral can change even though the lengths of the sides are fixed. In the case of the rack, the bolts only fix the lengths AB, BC, CD, and AD. (There is no "SSSS" congruence property for quadrilaterals.)

30. (a) The framework forms a parallelogram, but not necessarily a rectangle.

32. (a) Measures will vary, but $m(\angle AOB) = 2m(\angle APB)$.

   (b) $m(\angle AOB) = 2m(\angle APB)$

   (c) Draw the diameter $\overline{PQ}$. There are two cases to consider.
   Case 1:
   A and B lie on opposite sides of $\overline{PQ}$. We see that $\triangle POA$ is isosceles, so the base angles are congruent. That is, $m(\angle APO) = m(\angle OAP) = x$. Therefore, $m(\angle AOQ) = 2x = 2m(APO)$.

By the same reasoning, $m(BOQ) = 2y = 2m(\angle BPO)$. Thus $m(\angle AOB) = 2x + 2y = 2m(\angle APB)$.

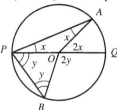

Case 2:
A and B lie on the same side of $\overline{PQ}$.
Again, $\triangle POA$ is isosceles, so $m(\angle APO) = m(\angle OAP) = x$. Therefore, $m(\angle AOQ) = 2x$. By the same reasoning, $m(BOQ) = 2y$. Thus, $m(\angle AOB) = 2x - 2y = 2m(\angle APB)$.

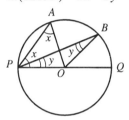

34. $m(\angle 3) = m(\angle 1) + m(\angle 2)$ (See Example 10.4.) Thus, $m(\angle 3) > m(\angle 1)$ and $m(\angle 3) > m(\angle 2)$.

JUST FOR FUN   Circle Amazement (page 728)

The jar lid will also give the circumscribing circle through A, B, C. When it is drawn, each of the four points A, B, C, and P is the intersection of three of the four circles.

Problem Set 11.2 (page 734)

1. (a) Step 1: Draw a line through P that intersects *l*. Label the intersection point A.
   Step 2: Draw arcs of equal radius centered at A and P. Label as B the intersection point of the arc at A with $\overleftrightarrow{AP}$. Label as C the intersection point of the arc at A with *l*. Label as D the intersection point of the arc at P with $\overleftrightarrow{AP}$.
   Step 3: Set the compass to radius BC, and draw the arc centered at D. Label as E the intersection with the arc drawn at P.
   Step 4: Construct the line k through P and E.

   (b) The construction gives the congruence $\triangle BAC \cong \triangle DPE$ by the SSS property. Thus the corresponding angles are congruent, $\angle BAC \cong \angle DPE$. Therefore $k \parallel l$ by the corresponding angles property.

3. (a) The corresponding angles property guarantees that *m* is parallel to *l*.

   (b) Align the ruler with line; slide the drafting triangle, with one leg of the right triangle on the ruler, until the second leg of the triangle meets point P.

   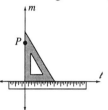

7. (a) The circumcenter will be inside the acute triangle.

   (b) The circumcenter will be at the midpoint of the hypotenuse of the right triangle.

   (c) The circumcenter will be outside the obtuse triangle.

   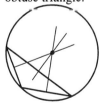

   (d) The circumcenter is inside, on, or outside a triangle if, and only if, the triangle is acute, right, or obtuse, respectively.

9. Since △PQS is inscribed in a circle with diameter $\overline{PQ}$, it has a right angle at point S by Thale's theorem. Thus $\overline{PS} \perp \overline{SQ}$. Similarly, △PQT is inscribed in a semi-circle with diameter $\overline{PQ}$, so $\overline{PT} \perp \overline{TQ}$. Hence $\overline{PS}$ and $\overline{PT}$ are tangent to the circle at Q.

12. By Thale's theorem, ∠ADB is a right angle. △ODB is an equilateral triangle since all sides have the length of the radius of the circles, so △ODB is equiangular by problem 8 of Problem Set 11.1. Thus, $m(\angle ODB) = 60°$. Moreover, ODBE is a rhombus, so $\overline{DE}$ is a bisector of ∠ODB, so $m(\angle EDB) = 30°$. Hence, $m(\angle ADE) = m(\angle ADB) - m(\angle EDB) = 90° - 30° = 60°$. Similarly, $m(\angle AED) = 60°$. Therefore, all angles of △ADE have measure 60°, so △ADE is equilateral by problem 9 of Problem Set 11.1.

14. Since
$m(\angle 1) + m(\angle 2) + m(\angle 3) + m(\angle 4) = 180°$,
$m(\angle 1) = m(\angle 2)$, and $m(\angle 3) = m(\angle 4)$,
$2m(\angle 2) + 2m(\angle 3) = 180°$.
Thus $m(\angle 2) + m(\angle 3) = 90°$. Hence, m and n are perpendicular.

15. (a) Suppose the perpendicular bisector of chord $\overline{AB}$ intersects the circle at a point C. Then the circle is the circumscribing circle of △ABC. The center of the circumscribing circle is the point of concurrence of the perpendicular bisectors of all three sides of △ABC. In particular, it is on the perpendicular bisector of side $\overline{AB}$ contains the center of the circle.

17. Constructions will vary.

    (a) Extend $\overline{AB}$, and construct the line at A that is perpendicular to $\overline{AB}$. Set the compass to radius AB and mark off this distance on the perpendicular line to determine a point C for which AC = AB. Similarly, construct a line through B perpendicular to $\overline{AB}$, and determine a point D (on the same side of $\overline{AB}$ as C) on this perpendicular so BD = AB. ABDC is a square with given side $\overline{AB}$.

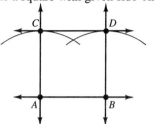

18. Constructions will vary.

    (a) Extend $\overline{AB}$. Construct perpendicular rays to $\overline{AB}$ at both A and B, to the same side of $\overline{AB}$. Bisect the right angle at A. Let the bisector's intersection with the ray at B determine C. Construct the line through C perpendicular to $\overline{BC}$. Let D be the intersection with the ray constructed at A. Thus ABCD is a square with the given side $\overline{AB}$.

21. By the Gauss-Wentzel Constructability Theorem, the constructible regular polygons have n sides where $n = 2^r \cdot 4$ (where r is a whole number), $2^r \cdot p$ (where p is a fermat prime and r is a whole number), or $n = 2^r \cdot p_1 \cdot \ldots \cdot p_m$ (where $p_1, \ldots p_m$ are distinct fermat primes and r is a whole number). The only fermat primes we need be concerned with are 3, 5, and 17, since the remaining fermat primes are greater than 100. Therefore, the polygon is constructible for the following values of n:
$2^r \cdot 4$: 4, 8, 16, 32, 64, 128, ...
$2^r \cdot 3$: 3, 6, 12, 24, 48, 96, 192, ...
$2^r \cdot 5$: 5, 10, 20, 40, 80, 160, ...
$2^r \cdot 17$: 17, 34, 68, 136, ...
$2^r \cdot 3 \cdot 5$: 15, 30, 60, 120, ...
$2^r \cdot 3 \cdot 17$: 51, 102, ...
$2^r \cdot 5 \cdot 17$: 85, 170, ...
$2^r \cdot 3 \cdot 5 \cdot 17$: 255, 510, ...
In summary we have the following list.
Constructible: 3, 4, 5, 6, 8, 10, 12, 15, 16, 17, 20, 24, 30, 32, 34, 40, 48, 51, 60, 64, 68, 80, 85, 96
Nonconstructible: 7, 9, 11, 13, 14, 18, 19, 21, 22, 23, 25, 26, 27, 28, 29, 31, 33, 35, 36, 37, 38, 39, 41, 42, 43, 44, 45, 46, 47, 49, 50, 52, 53, 54, 55, 56, 57, 58, 59, 61, 62, 63, 65, 66, 67, 69, 70, 71, 72, 73, 74, 75, 76, 78, 79, 81, 82, 83, 84, 86, 87, 88, 89, 90, 91, 92, 93, 94, 95, 97, 98, 99, 100

23. $\triangle ABD \cong \triangle ACD$ by the SSS property. Therefore, $\angle BAD \cong \angle CAD$.

25. Sketches will vary.

    (a) $ABDC$ is a parallelogram, so $AB = CD$ and $BD = AC$. (See problem 11(b) of Problem Set 11.1.) By the SSS property, $\triangle ABC \cong \triangle DCB$. Similarly, using parallelograms $ACBF$ and $ABCE$, $\triangle ABC \cong \triangle BAF$ and $\triangle ABC \cong \triangle CEA$.

    (c) The altitude $\triangle ABC$ at A is perpendicular to $\overline{BC}$. But $\overline{FE}$, by construction, is parallel to $\overline{BC}$ so the altitude at A is also perpendicular to $\overline{FE}$. Since A is the midpoint of $\overline{FE}$, this means the altitude of $\triangle ABC$ at A is also the perpendicular bisector of $\overline{FE}$. The same reasoning applies to the altitudes at B and C.

27. $m(\angle NOQ) = m(\angle PON) - m(\angle POQ)$ $= 144° - 120° = 24°$ since a central angle for a regular pentagon is $72°$ and a central angle for an equilateral triangle is $120°$. A regular 15-gon has central angle $\frac{360°}{15} = 24°$, so laying off segments of length QN would give 15 equally spaced points around the circle.

30. Use the laws of exponents and the distributive property of multiplication over addition to show that $F_5 = 2^{2^5} + 1 = 2^{32} + 1$
$= 2^{12+20} + 1 = 2^{12} \cdot 2^{20} + 1$
$= (4096) \cdot (1,048,576) + 1$
$= 4 \cdot (1,048,576) \cdot 10^3 + 96 \cdot (10^6 + 48,576) + 1$
$= 4,194,304,000 + 96,000,000 + 4,663,296 + 1$
$= 4,294,967,297$.

32. (a) G, H, and P are collinear. The Euler line passes through G, H, and P.

    (b) $\frac{GH}{PG} = 2$. Thus G is one third of the distance from P to H along the Euler line.

    (c) The circle intersects all sides of $\triangle ABC$ at their midpoints.

    (d) The circle bisects each of the segments $\overline{AH}$, $\overline{BH}$, and $\overline{CH}$.

34. $\triangle TRI$ is equilateral.

36. (a) True. $\angle A \cong \angle D$, $\angle B \cong \angle E$, $\angle C \cong \angle F$, $\overline{AB} \cong \overline{DF}$, and $\overline{BC} \cong \overline{DE}$.

    (b) False. After pairing up the congruent angles, the sides with equal lengths are not corresponding sides in the triangles, so the two triangles are not congruent.

JUST FOR FUN Thales' Puzzle (page 745)

Later in the day, the points A and C will cast new shadows, say at P' and Q'. Since both P and P' are away from the pyramid, Thales can easily measure PP' and QQ', and calculate the scale factor $\frac{PP'}{QQ'} = s$. The height of the pyramid is therefore sh, where h is the height of the vertical stick.

Problem Set 11.3 (page 748)

1. (a) $m(\angle A) = 180° - m(\angle B) - m(\angle C)$ $= 180° - 60° - 90° = 30°$. $\triangle ABC \sim \triangle PNO$ by the AA similarity property. The scale factor from $\triangle ABC$ to $\triangle PNO$ is $\frac{PN}{AB} = \frac{12}{8} = \frac{3}{2}$.

   (c) $\angle HIG \cong \angle UIT$ since they are vertical angles. $\triangle GHI \sim \triangle TUI$ by the AA similarity property. The scale factor is $\frac{UT}{HG} = \frac{8}{5}$.

2. (a) Yes; since all angles measure $60°$, equilateral triangles are similar by the AA similarity property.

   (c) Yes; since any two such triangles have angles of measure $36°$ and $90°$, they are similar by the AA similarity property.

   (e) Yes; any two congruent triangles are similar by the AA similarity property, the SSS similarity property, or the SAS similarity property. The scale factor is 1.

3. (a) $\frac{15}{12} = \frac{a}{8}$; $a = \frac{15}{12} \cdot 8 = 10$

   (c) $\frac{15+3}{15} = \frac{c+2}{c}$; $18c = 15c + 30$; $c = 10$

4. (a) No; for example, a square and a nonsquare rectangle are convex quadrilaterals with congruent angles, yet they are not similar.

6. (a) $\overline{AB} \parallel \overline{CD}$ so $\angle ABE \cong \angle CDE$ by the alternate interior angles theorem. $\angle AEB \cong \angle CED$ since they are vertical angles. $\triangle ABE \sim \triangle CDE$ by the AA similarity property.

(b) By similarity, $\frac{x}{36} = \frac{17}{51}$, so $x = \frac{17}{51} \cdot 36 = 12$. Also, $\frac{26}{y} = \frac{17}{51}$, so $y = 26 \cdot \frac{51}{17} = 78$.

8. (a) $\angle CAD \cong \angle BAC$, since they are the same angle and $m(\angle ADC) = m(\angle ACB) = 90°$. By the AA similarity property, $\triangle ADC \sim \triangle ACB$. Likewise, $\triangle CDB \sim \triangle ACB$. Thus $\triangle ADC \sim \triangle CDB$.

10. (a) Draw an arc of large enough radius so that point B is on the seventh line above the line with point A.

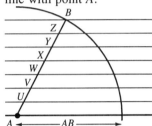

11. (a) Since corresponding angles of parallel lines are congruent, $\angle OXU \cong \angle OAB$ and $\angle OUX \cong \angle OBA$. By the AA similarity property, $\triangle OXU \sim \triangle OAB$. Thus $\frac{x}{a} = \frac{1}{b}$ or $x = \frac{a}{b}$.

(b) Construct a triangle $\triangle OAB$ with $OA = a$ and $OB = b$. Mark the point U at length 1 from point O on segment $\overline{OB}$ or its extension. Construct a line through U which is parallel to $\overline{AB}$, calling the point where it crosses $\overline{OA}$ the point X. Then $OX = \frac{a}{b}$ by the reasoning in part (a).

13. (a) $\angle ACB \cong \angle CDA$ since they are right angles. $\angle BAC \cong \angle CAD$ since they are the same angle. $\triangle ABC \sim \triangle ACD$ by the AA similarity property.
$\angle BCA \cong \angle BDC$ since they are right angles. $\angle ABC \cong \angle CBD$ since they are the same angle. $\triangle ABC \sim \triangle CBD$ by the AA similarity property.

(b) Using $\triangle ACD \sim \triangle ABC$, then $\frac{AD}{AC} = \frac{AC}{AB}$ or $\frac{x}{b} = \frac{b}{c}$. Similarly, $\triangle CBD \sim \triangle ABC$ gives $\frac{BD}{BC} = \frac{CB}{AB}$ or $\frac{y}{a} = \frac{a}{c}$.

15. (a) $\angle APC \cong \angle BAD$ since corresponding angles of parallel lines are congruent. $\angle BAD \cong \angle DAC$ since $\overline{AD}$ bisects $\angle BAC$. $\angle DAC \cong \angle ACP$ by the alternate interior angles theorem. Thus $\angle APC \cong \angle BAD \cong \angle DAC \cong \angle ACP$.

17. (a) Since the result of Example 11.14 can be applied to space quadrilaterals, the midpoints W, X, Y, and Z are the vertices of the parallelogram WXYZ, and we know that the diagonals of any parallelogram intersect at their common midpoints.

(b) By using part (a) on PQRS, $\overline{WY}$ and $\overline{XZ}$ intersect at their common midpoint. Applying the reasoning in part (a) on the quadrilateral PSQR, $\overline{XZ}$ and $\overline{UV}$ intersect at their common midpoint. Thus, all three bimedians intersect at their common midpoints.

19. Construct $\overline{AC}$. Let L denote the midpoint of $\overline{AC}$, which is also the midpoint of $\overline{BD}$. By Example 11.15, P is the centroid of $\triangle ABC$ and $BP = \frac{2}{3} BL$. Since $BL = \frac{1}{2} BD$, this shows $BP = \frac{2}{3} \cdot \frac{1}{2} BD = \frac{1}{3} BD$. By the same reasoning, Q is the centroid of $\triangle ADC$ and $QD = \frac{1}{3} BD$. Finally, $PQ = BD - BP - BD = \left(1 - \frac{1}{3} - \frac{1}{3}\right) BD = \frac{1}{3} BD$. Therefore, $BP = PQ = QD$.

21. $\angle APC \cong \angle BPD$ since they are vertical angles. $\angle PDB \cong \angle PCA$ since they are right angles. $\triangle ACP \sim \triangle BDP$ by the AA similarity property. Since $\frac{AC}{BD} = \frac{4}{2} = 2$, the scale factor is 2. Therefore, $CP = 2DP$ and $AP = 2BP$. Since $CD = 4$ and $CD = CP + DP = 3DP$, $DP = \frac{4}{3}$, and $CP = \frac{8}{3}$. Using the Pythagorean theorem, $BP = \sqrt{2^2 + \left(\frac{4}{3}\right)^2} = \sqrt{4 + \frac{16}{9}} = \sqrt{\frac{52}{9}} = \frac{2}{3}\sqrt{13}$ and $AP = 2BP = \frac{4}{3}\sqrt{13}$.

23. The right triangles also have a congruent angle at the vertex at the mirror, so the triangles are similar by the AA property. Assuming Mohini's eyes are 5" beneath the top of her head, this gives the proportion $\frac{h}{5'} = \frac{15'}{4'}$,

making the pole $h = (5')\left(\dfrac{15'}{4'}\right) = 18'9''$ high.

Mohini must take into account how far from the ground her eyes are.

25. By similar triangles $\dfrac{6' - x}{6'} = \dfrac{5.25'}{18'}$. Therefore, $x = 6'\left(1 - \dfrac{5.25'}{18'}\right) = 4.25' = 4'3''$.

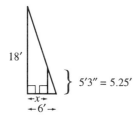

27. Drawings will vary.

   (a) $\triangle BCP \sim \triangle DAP$

   (b) Observe that $PA \cdot PB = PC \cdot PD$; part (a) gives $\dfrac{PA}{PC} = \dfrac{PD}{PB}$ or $PA \cdot PB = PC \cdot PD$ so, if proven, it would justify our observation.

30. $\triangle ABP \cong \triangle DPC$ by the SSS property; $\triangle ABC \cong \triangle DCB$ by the SSS property since $AC = AP + PC = DP + PB = DB$, $AB = DC$, and $BC = CB$; $\triangle ADB \cong \triangle DAC$ by the SSS property since $AC = DB$, $DC = AB$, and $AD = DA$.

32. Draw $\overline{PC}$. $\triangle PAC$ and $\triangle PBC$ are right triangles. $AC \cong BC$ and $PC \cong PC$. By the HL theorem, $\triangle ACP \cong \triangle BCP$, so $PA = PB$.

CLASSIC CONUNDRUM Diagonal Diagnosis (page 754)

The radius is $5 + 4 = 9$. Since $ABCD$ is a rectangle, $AC = DB$. $\overline{DB}$ is a radius, so $AC = 9$.

Chapter Review Exercises (page 756)

1. (a) $\overline{AD} \cong \overline{AB}$, $\overline{CD} \cong \overline{CB}$ and $\overline{AC} \cong \overline{AC}$. $\triangle ACD \cong \triangle ACB$ the SSS congruence property.

   (b) $\overline{AC} \cong \overline{EC}$, $\overline{CD} \cong \overline{CB}$, and $\angle C \cong \angle C$. $\triangle ACD \cong \triangle ECB$ by the SAS congruence property.

   (c) $\angle A \cong \angle B$, $\angle F \cong \angle C$, and $\overline{FD} \cong \overline{CE}$. $\triangle ADF \cong \triangle BEC$ by the AAS congruence property.

   (d) $\overline{CA} \cong \overline{CE}$, $\overline{CD} \cong \overline{CB}$, and $\angle C \cong \angle C$. $\triangle ACD \cong \triangle ECB$ by the SAS congruence property.

   (e) $\triangle ABE$ is isosceles, so $\angle B \cong \angle E$ by the isosceles triangle theorem. $\angle BAC \cong \angle EAD$ and $\overline{BA} \cong \overline{EA}$. $\triangle ABC \cong \triangle AED$ by the ASA congruence theorem. (It is also true that $\triangle ABD \cong \triangle AEC$.)

   (f) $\triangle ABC$ and $\triangle ADC$ are right triangles. $\overline{AC} \cong \overline{AC}$ and $\overline{AB} \cong \overline{AD}$. $\triangle ABC \cong \triangle ADC$ by the HL theorem.

2. $m(\angle B) = m(\angle G) = 62°$, $AB = FG = 2.1$ cm, and $BC = GH = 3.2$ cm, so $\triangle ABC \cong \triangle FGH$ by the SAS congruence property.

   (a) $AC = FH = 2.9$ cm

   (b) $m(\angle H) = 180° - m(\angle F) - m(\angle G) = 180° - 78° - 62° = 40°$

   (c) $m(\angle A) = m(\angle F) = 78°$

   (d) $m(\angle C) = m(\angle H) = 40°$

3. $\angle B \cong \angle C$ by the isosceles triangle theorem. By construction, $BF = CD$ and $BD = CE$. Therefore, $\triangle BDF \cong \triangle CED$ by the SAS congruence property, so $DE = DF$.

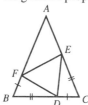

4. (a) See construction 6 of Section 11.2.

   (b) See construction 2 of Section 11.2.

   (c) See construction 5 of Section 11.2.

   (d) One method is given in the back of the textbook. Alternately, fix a point A on line m. Use construction 4 to construct a perpendicular line through A. The line is also perpendicular to ℓ. Let the intersection of the line and ℓ be point B. Use construction 5 to construct the perpendicular bisector of $\overline{AB}$. This line is k.

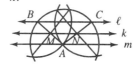

5. (a) Reflect one side of ∠A to the other.

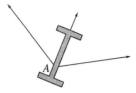

(b) Pivot the Mira about P until the line reflects onto itself.

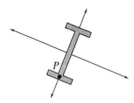

(c) Reflect A onto B.

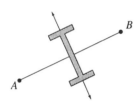

(d) Reflect m onto line ℓ.

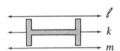

6. (a) Construct ∠A, mark off length AB, and draw a circle at B of radius BC. The circle intersects the other ray from A at two points, C_1 and C_2, giving two triangles $\triangle ABC_1$ and $\triangle ABC_2$.

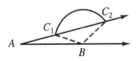

(b) Only $\triangle ABC_1$ has $\angle C = \angle C_1$ obtuse.

7. Find the midpoint M of $\overline{AD}$. Then draw circles of radius $\frac{1}{2}AD$ centered at A, D, and M. $AB = CD = DE = FA = \frac{1}{2}AD$ by the construction. Since $\triangle BCM$ and $\triangle EFM$ are equilateral, $BC = EF = \frac{1}{2}AD$. Thus, ABCDEF is a regular hexagon.

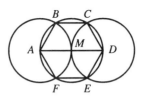

8. (a) Yes; measure the side lengths and see if the SSS similarity property applies.

(b) Yes; measure the angles and see if the AA similarity property applies.

9. Bisect sides $\overline{AB}$ and $\overline{AC}$ to determine their midpoints, M and N, respectively. Extend $\overline{AB}$ beyond B and draw the circle at B through M. Let E be the intersection with the extension. Similarly, extend $\overline{AC}$ beyond C, draw the circle at C through N, and let F be the intersection of this circle with the extension. Choose D = A, so $\angle A \cong \angle D$. Since
$AM = MB = BE = \frac{1}{2}AB$ and
$AN = NC = CF = \frac{1}{2}AC$,
$DE = AE = AM + MB + BE = \frac{3}{2}AB$ and
$DF = AF = AN + NC + CF = \frac{3}{2}AC$. Thus,
$\frac{DE}{AB} = \frac{DF}{AC} = \frac{3}{2}$. $\triangle ABC \sim \triangle DEF$ by the SAS similarity property.

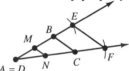

10. (a) $\angle A \cong \angle Q$ and $\frac{QP}{AB} = \frac{QR}{AC} = \frac{3}{2}$, so $\triangle ABC \sim \triangle QPR$ by the SAS similarity property. The scale factor is $\frac{3}{2}$.

(b) $\frac{XZ}{CB} = \frac{XY}{CA} = \frac{YZ}{AB} = 3$, so $\triangle ABC \sim \triangle YZX$ by the SSS similarity property. The scale factor is 3.

(c) $m(\angle F) = m(\angle H) = 70°$ by using the isosceles triangle theorem. $\triangle ABC \sim \triangle HGF$ by the AA similarity property. The scale factor is $\frac{HG}{AB} = \frac{4}{6} = \frac{2}{3}$.

(d) $\frac{BD}{AB} = \frac{DC}{BD} = \frac{BC}{AD} = \frac{1}{2}$. $\triangle ABD \sim \triangle BDC$ by the SSS similarity property. The scale factor is $\frac{1}{2}$.

11. One method is given in the back of the textbook. Another way is to notice that, since k, l, and m are parallel, the small, medium, and large triangles are similar.
$\frac{x}{x+9} = \frac{16}{16+12}$ or
$28x = 16x + 144$. Thus $12x = 144$ or $x = 12$.
$\frac{16}{16+12+y} = \frac{x}{x+9+15}$ or $\frac{16}{28+y} = \frac{12}{36}$.
Thus $y + 28 = 48$ or $y = 20$.

Chapter 11 Test (page 758)

1. (a) $\overline{AC} \cong \overline{AC}$, $\angle ACD \cong \angle ACB$, and $\angle D \cong \angle B$. $\triangle ACD \cong \triangle ACB$ by the AAS congruence property.

   (b) $\overline{AB} \cong \overline{AD}$, $\overline{AC} \cong \overline{AC}$, and $\angle BAC \cong \angle DAC$. $\triangle ABC \cong \triangle ADC$ by the SAS congruence property.

   (c) $m(\angle ADC) = m(\angle ADB) + m(\angle BDC)$
   $= m(\angle BCA) + m(\angle ACD) = m(\angle BCD)$, so $\angle ADC \cong \angle BCD$. $\angle ACD \cong \angle BDC$ and $\overline{DC} \cong \overline{CD}$. $\triangle ADC \cong \triangle BCD$ by the ASA congruence property.

   (d) $\angle ABE \cong \angle CBD$ since they are vertical angles. $\angle A \cong \angle C$ and $\overline{AE} \cong \overline{CD}$. $\triangle ABE \cong \triangle CBD$ by the AAS congruence property.

   (e) $\angle BDC \cong \angle FDE$ since they are vertical angles. $\angle DBC \cong \angle DFE$ since they are right angles. $\overline{BD} \cong \overline{FD}$. $\triangle BCD \cong \triangle FED$ by the ASA congruence property. (Note that $\triangle ACF \cong \triangle AEB$ also.)

   (f) $\overline{AB} \cong \overline{AD}$, $\overline{BC} \cong \overline{DC}$, and $\overline{AC} \cong \overline{AC}$. $\triangle ABC \cong \triangle ADC$ by the SSS congruence property.

2. Let x be the length of third side in feet. Use the triangle inequality. $x < 10 + 16$, so $x < 26$. $16 < 10 + x$, so $6 < x$. Thus the third side is greater than 6 feet and less than 26 feet.

3. Construct a segment $\overline{DE}$ that is congruent to $\overline{AB}$. Construct rays at D and E to the same side of $\overline{DE}$ to form angles that are respectively congruent to $\angle A$ and $\angle B$. Let F be a point of intersection of the rays. Then $\triangle DEF \cong \triangle ABC$.

4. (a) $\triangle ABC$ and $\triangle DEF$ are right triangles. $BC = EF$ and $AC = DF$. $\triangle ABC \cong \triangle DEF$ by the HL theorem.

   (b) They are not congruent. The hypotenuse of $\triangle ABC$ is 5. Since $EF = 5$, the hypotenuse of $\triangle DEF$ must have length greater than 5, so the hypotenuses cannot correspond.

   (c) $\triangle ABD$ and $\triangle CBD$ are right triangles. $BD = BD$ and $AB = CB$. $\triangle ABD \cong \triangle CBD$ by the HL theorem.

   (d) $\angle ABE \cong \angle DBC$ since they are right triangles. $AB = DB$ and $BE = BC$. $\triangle ABE \cong \triangle DBC$ by the SAS property.

5. Since $ABCDE$ is a regular pentagon, $\angle A \cong \angle B \cong \angle C \cong \angle D \cong \angle E$ and $AB = BC = CD = DE = EA$. Since $AP = BQ = CR = DS = ET$ is given, $PB = QC = RD = SE = TA$. Thus $\triangle APT \cong \triangle BQP \cong \triangle CRQ \cong \triangle DSR \cong \triangle ETS$ by the SAS property, so $PT = QP = RQ = SR = TS$. Therefore, $PQRST$ is equilateral. If $m(\angle APT) = x$ and $m(\angle ATP) = y$, then the measure of all the interior angles of $PQRST$ is $180° - x - y$. Thus, $PQRST$ is regular.

6. Since $\angle F \cong \angle G$, $FH = GH$ by the isosceles triangle theorem. Since $FG = FH$, $\triangle FGH$ is equilateral.

7. (a) Draw circles of radius PQ, one centered at P and one centered at Q. The circles intersect at points R and S for which $\triangle PQR$ and $\triangle PQS$ are equilateral.

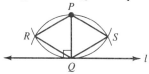

(b) Using R and S from part (a), construct the angle bisectors of $\angle QPR$ and $\angle QPS$, and denote their intersections with line l as T and U. $m(\angle QPR) = m(\angle QPS) = 60°$, so $m(\angle QPT) = m(\angle QPU) = 30°$. Thus $m(\angle TPU) = 60°$. $m(\angle PTQ) = m(\angle PUQ) = 180° - 30° - 90° = 60°$. Therefore, $\triangle PTU$ is equiangular. Hence, $\triangle PTU$ is the desired equilateral triangle.

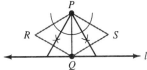

8. (a) $\triangle ADE \sim \triangle ACB$ by the AA similarity property, since both triangles contain $\angle A$ and a right angle.

   (b) $\triangle ABC \sim \triangle XYZ$ by the SAS similarity property since $\frac{XY}{AB} = \frac{YZ}{BC} = \frac{5}{4}$ and $\angle B \cong \angle Y$.

   (c) $\triangle DEG \sim \triangle EFG$ by the SSS similarity property, since $\frac{EF}{DE} = \frac{FG}{EG} = \frac{GE}{GD} = 2$.

   (d) $\triangle AEB \sim \triangle CED$ by the AA similarity property since $\angle AEB \cong \angle CED$ being vertical angles and $\angle EBA \cong \angle EDC$ being alternate interior angles between parallel lines.

9. (a) $m(\angle W) = m(\angle M) = 53°$

   (b) Scale factor $= \frac{WU}{MK} = \frac{20}{16} = \frac{5}{4}$

   (c) $\frac{UV}{KL} = \frac{5}{4}$, so $UV = \frac{5}{4} KL = \frac{5}{4} \cdot 20 = 25$.

10. (a) Make the assumption that they form the same angle relative to the ground, namely the angle of elevation, since the distant sun casts shadow rays that are essentially parallel.

    (b) Make the assumption that the person and the tree stand at the same angle with the ground, for example, both vertical. Then $\triangle ABC \sim \triangle DEF$ by the AA similarity property.

    (c) $\frac{DE}{AB} = \frac{DF}{AC}$, so $DE = \frac{56'}{7'} \cdot 6' = 48'$.
    The tree is 48 ft tall.

11. $\angle A \cong \angle A$ and $\angle ADE \cong \angle ACB$ since they are right angles. $\triangle ADE \sim \triangle ACB$ by the AA similarity property. Thus
$\frac{AE}{AB} = \frac{AD}{AC} = \frac{AD}{AD+DC} = \frac{2DC}{2DC+DC} = \frac{2}{3}$, so
$AE = \frac{2}{3} AB = \frac{2}{3} \cdot 12 = 8$.
$EB = AB - AE = 12 - 8 = 4$.

12. Draw the circle with radius MC centered at M. The circle passes through A, B, and C. Thus, by Thale's theorem, $\triangle ABC$ is a right triangle because it is inscribed in a semicircle of diameter $\overline{AB}$.

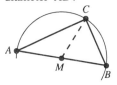

Chapter 12

Problem Set 12.1 (page 773)

1. (a) Height, length, thickness, area, diagonal, weight

   (c) Height, length, depth

3. (a) Answers will vary anywhere from 40 to 54.

   (b) The circular portions of area do not fit well together, and leave gaps between them.

5. (a) $1 \text{ acre} = \frac{1}{640} \text{ mi}^2 = \left(\frac{1}{640} \text{ mi}^2\right)\left(\frac{5280 \text{ ft}}{1 \text{ mi}}\right)^2 = 43{,}560 \text{ ft}^2$

6. (a) 3 in

7. (a) The true distance is between 165.5 km and 166.5 km.

   (c) The true diameter is between 0.495 mm and 0.505 mm.

9. (a) $33 \text{ cL} = 33 \times 10^{-2} \text{ L} = 330 \times 10^{-3} \text{ L} = 330 \text{ mL}$

   (b) Not quite; 1 L = 100 mL and three bottles has volume $3 \times 33 \text{ cL} = 990 \text{ mL}$.

10. (a) $58{,}728 \text{ g} = 58.728 \times 10^3 \text{ g} = 58.728 \text{ kg}$

    (c) $0.23 \text{ kg} = 0.23 \times 10^3 \text{ g} = 230 \text{ g}$

11. (a) 3.5 kg

13. (a) About 27.9 cm by 21.6 cm

    (c) About 2 cm

15. (a) 1 jill = 2 jacks = 4 jiggers = 8 mouthfuls;
    1 cup = 2 jills = 16 mouthfuls;
    1 pint = 2 cups = 32 mouthfuls

17. $1 \text{ ha} \doteq (10{,}000 \text{ m}^2)\left(\frac{1 \text{ km}}{1000 \text{ m}}\right)^2\left(\frac{1 \text{ mi}}{1.6 \text{ km}}\right)^2\left(\frac{640 \text{ acre}}{1 \text{ mi}^2}\right) = 2.5 \text{ acres}$

19. $\left(\frac{25 \text{ in}}{1 \text{ min}}\right)\left(\frac{60 \text{ min}}{1 \text{ hr}}\right)\left(\frac{24 \text{ hr}}{1 \text{ day}}\right)\left(\frac{14 \text{ day}}{1 \text{ fortnight}}\right)\left(\frac{1 \text{ ft}}{12 \text{ in}}\right)\left(\frac{1 \text{ furlong}}{660 \text{ ft}}\right) \doteq 63.6 \text{ furlong/fortnight}$

21. (a) kilo

    (c) no prefix

    (e) milli

23. $\left(\frac{100 \text{ km}}{9 \text{ L}}\right)\left(\frac{3.7854 \text{ L}}{1 \text{ gal}}\right)\left(\frac{1 \text{ mi}}{1.6 \text{ km}}\right) \doteq 26.3 \text{ mi/gal}$

25. (a) $(5 \text{ gal})\left(\frac{4 \text{ qt}}{1 \text{ gal}}\right)\left(\frac{32 \text{ oz}}{1 \text{ qt}}\right) = 640 \text{ oz}$

    Since $\frac{640}{80} = 8$, add 8 liquid ounces of concentrate.

    (b) Add $80 \times 65 \text{ mL} = 5200 \text{ mL} = 5.2 \text{ L}$ of water.

27. A league varied from time to time in history, but was usually close to 3 miles. Thus the Nautilus traveled about 60,000 miles.

31. Interior angle measures of a triangle sum to 180°, so $8x + 6x + 4x = 180°$, or $18x = 180°$. Thus $x = 10°$. The angles measure 80°, 60°, and 40°.

33. (a) $\triangle ABC \cong \triangle EFD$ by the SAS congruence property.

    (c) $\frac{EF}{AB} = \frac{ED}{BC} = 3$;
    $m(\angle E) = 180° - m(\angle D) - m(\angle F)$
    $= 35° = m(\angle B)$; $\triangle ABC \sim \triangle FED$ by the SAS similarity property.

JUST FOR FUN   **How to Cover a Long Hole with a Short Board (page 778)**

JUST FOR FUN   **Tile and Smile (page 781)**

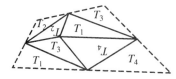

The parallelogram has one-half of the area of the quadrilateral

Problem Set 12.2 (page 789)

1. (a) A 2-meter square has area $4\ m^2$, which is larger than $2\ m^2$.

   (b) A square of area $\frac{1}{2}\ m^2$ is larger than the area, $\frac{1}{4}\ m^2$, of a square with sides $\frac{1}{2}$ m.

3. (a) 12 units

4. Rearrange the triangles numbered 1–9 in the gap between the 12-gon and the square, so $\frac{3}{4}$ of the square is covered. Hence the area of the 12-gon is $\frac{3}{4}$ (4 units) = 3 units of area.

6. (a) $36\ cm^2 \div 3\ cm = 12\ cm$

7. (a) 1 cm by 24 cm; 2 cm by 12 cm; 3 cm by 8 cm; 4 cm by 6 cm. The dimensions can also be given in opposite order.

9. (a) Starting with the two triangles on top of each other, rotate the top one 180° about the midpoint of any side to form a parallelogram.

   (b) Two triangles combine to form a parallelogram with base length the base of the triangle and height the height of the triangle. Therefore
   $2 \cdot$ area (triangle)
   = area (parallelogram) = bh, so
   area (triangle) = $\frac{1}{2}bh$.

11. (a) Put two copies of the trapezoid together as shown.

    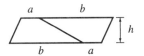

    $2 \cdot$ area(trapezoid)
    = area(parallelogram)
    = $h(a + b)$,
    so area(trapezoid) = $\frac{1}{2}(a + b)h$

12. (a) $A = (16.2\ ft)(5\ ft) = 81\ ft^2$;
    $P = 2(16.2\ ft) + 2(6.4\ ft) = 45.2\ ft$

13. (a) $A = \frac{1}{2}(81.2\ m)(41.0\ m) = 1664.6\ m^2$;
    $P = 49.0\ m + 68.1\ m + 81.2\ m$
    $= 198.3\ m$

14. (a) $A = \frac{1}{2}(19 + 25)10 = 220$ square units

15. (a) $\triangle ABC$; all triangles have the same base and $\triangle ABC$ has the smallest height.

    (b) $\triangle ABF$; it has the largest height.

    (c) $\triangle ABD$ and $\triangle ABE$; they have equal heights and the same base.

17. (a) 9 square units

19. **(a)** $100 \text{ m} + 100 \text{ m} + 2\pi \cdot 25 \text{ m} = (200 + 50\pi)\text{m} \doteq 357 \text{ m}$

   (b) $(50 \text{ m})(100 \text{ m}) + \pi(25 \text{ m})^2 = (5000 + 625\pi)\text{m}^2 \doteq 6963 \text{ m}^2$

20. **(a)** $\pi(2)^2 - \pi(1)^2 = 3\pi$ square units $\doteq 9.4$ square units

22. **(a)** Since the distance from the North Pole to the equator is one-quarter the circumference, the circumference is
   $(4)(10,000,000)(1 \text{ m}) = 40,000,000 \text{ m}$.

   (b) $12,755\pi \text{ km} \doteq 40,071 \text{ km} = 40,071,000 \text{ m}$

23. The common overlap reduces the area of both regions by the same amount, so the difference in area is unchanged. Thus the area is 20 cm².

25. **(a)**

   The area of the unshaded region in the above figure is
   $1^2 - \frac{1}{4}\pi(1^2) = 1 - \frac{\pi}{4}$ square units.
   There are two such unshaded regions in the figure below.

   The area of the shaded region in the figure above is $1 - \left(1 - \frac{\pi}{4}\right) - \left(1 - \frac{\pi}{4}\right) = \frac{\pi}{2} - 1$ square units.

27. The areas of the rectangular portions of sidewalk total
   $(60 \text{ ft} + 70 \text{ ft} + 40 \text{ ft} + 80 \text{ ft} + 50 \text{ ft})(8 \text{ ft}) = 2400 \text{ ft}^2$.
   The pieces formed with circular areas have total turning of 360°, so when placed together would form a circle with radius 8 ft, and area of $\pi(8 \text{ ft})^2 = 64\pi \text{ ft}^2$. Total area is $(2400 + 64\pi)\text{ft}^2 \doteq 2601 \text{ ft}^2$.

29. **(a)** The board has moved the length of twice the circumference, or 20″.

30. **(a)** The areas are given in order from left to right. $\frac{1}{2}\pi(2^2) = 2\pi$; $\frac{1}{4}\pi(3^2) = \frac{9}{4}\pi$;
   $\frac{1}{6}\pi(5^2) = \frac{25}{6}\pi$; $\frac{1}{360}\pi(4^2) = \frac{16}{360}\pi$

   (b) $\left(\frac{x}{360}\right)(\pi r^2)$

32. Draw $\overline{AP}$, $\overline{BP}$, and $\overline{CP}$. Then area($\triangle ABC$) = $\frac{1}{2}sh$
   = area($\triangle ABP$) + area($\triangle BPC$)
      + are($\triangle CPA$)
   = $\frac{1}{2}sx + \frac{1}{2}sz + \frac{1}{2}sy = \frac{1}{2}s(x + y + z)$. Therefore, $\frac{1}{2}sh = \frac{1}{2}s(x + y + z)$, so
   $h = x + y + z$.
   Alternate visual proof:

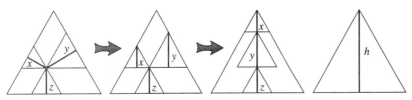

35. Consider the carpet as a 6 ft by 4 ft rectangle with two semicircular ends of radius 2 ft. Then the carpet's area is
$(6 \text{ ft})(4 \text{ ft}) + \pi(2 \text{ ft})^2 \doteq 36.6 \text{ ft}^2 \doteq 5266 \text{ in}^2$,
so about 5266 in. of braid, or about 439 ft.

37. area(kitchen) = $(10 \text{ ft})(12 \text{ ft})$
    $= (120 \text{ ft}^2)\left(\dfrac{144 \text{ in}^2}{1 \text{ ft}^2}\right) = 17,280 \text{ in}^2$, so the
    number of tiles needed is $\dfrac{17,280 \text{ in}^2}{64 \text{ in}^2} = 270$.
    Some extra tiles should also be ordered to account for mistakes, waste, and so on.

39. View one side of length 300 ft as a base. The altitude of the triangle is greatest if the angle is 90°.
    Thus $m(\angle A) = 90°$.

40. (a) $s = \frac{1}{2}(10 + 10 + 12) = 16$, so
    $A = \sqrt{16(16-10)(16-10)(16-12)} = 48$ square units

42. The circumscribed circle has four times the area of the inscribed circle.

44. (a) $1 \text{ m} = 100 \times 10^{-2} \text{ m} = 100 \text{ cm}$

    (c) $1 \text{ m}^2 = (100 \text{ cm})^2 = 10,000 \text{ cm}^2$

46. For rectangle $ABCD$, $\overline{AB} \cong \overline{DC}$, $\overline{BC} \cong \overline{CB}$, and $\angle ABC \cong \angle DCB$ since they are right angles. Thus, $\triangle ABC \cong \triangle DCB$ by the SAS property. Therefore, $\overline{AC} \cong \overline{BD}$, so the diagonals are congruent.

Problem Set 12.3 (page 801)

1. (a) By the Pythagorean theorem,
    $x^2 = 7^2 + 24^2 = 49 + 576 = 625$.
    Therefore, $x = \sqrt{625} = 25$.

   (c) By the Pythagorean theorem,
    $x^2 + 5^2 = 22^2$ or
    $x^2 = 22^2 - 5^2 = 484 - 25 = 459$.
    Therefore, $x = \sqrt{459}$.

   (e) By the Pythagorean theorem,
    $\left(\dfrac{x}{2}\right)^2 + 1 = x^2$ or $\left(\dfrac{3}{4}\right)x^2 = 1$ or $x^2 = \dfrac{4}{3}$.
    Therefore, $x = \dfrac{2}{\sqrt{3}}$ (or $\dfrac{2\sqrt{3}}{3}$).

2. (a) By the Pythagorean theorem,
    $x^2 + (2x)^2 = (25)^2$ or $5x^2 = 625$.
    Therefore, $x = \sqrt{125} = 5\sqrt{5}$.

3. (a) $x^2 = 10^2 + 15^2 = 325$, so $x = \sqrt{325}$;
    $y^2 = x^2 + 7^2 = 325 + 49 = 374$, so
    $y = \sqrt{374}$.

4. (a) $x^2 + 12^2 = 13^2$ or $x^2 = 25$, so
    $x = \sqrt{25} = 5$.

5. (a) height = $\sqrt{15^2 - 9^2} = 12$, so
    area = $(20)(12) = 240$ square units.

7. The smaller circle has radius 1 and area $\pi(1^2) = \pi$ square units. By the Pythagorean theorem, the larger circle has radius $\sqrt{1^2 + 1^2} = \sqrt{2}$ and area $\pi\left(\sqrt{2}\right)^2 = 2\pi$ square units.
    Thus, the area between the two circles is $2\pi - \pi = \pi$ square units.
    Therefore, the areas are equal.

9. $AC = \sqrt{4+1} = \sqrt{5}$; $AD = \sqrt{5+1} = \sqrt{6}$;
    $AE = \sqrt{6+1} = \sqrt{7}$; $AF = \sqrt{7+1} = \sqrt{8}$;
    $AG = \sqrt{8+1} = \sqrt{9} = 3$

11. Consider the following figure.

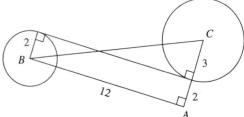

The distance between centers, B and C, is $\sqrt{12^2 + 5^2} = 13$.

13. (a) $(21)^2 + (28)^2 = 1225 = (35)^2$; yes, they can.

    (c) $(12)^2 + (35)^2 = 1369 = (37)^2$; yes, they can.

    (e) $(7\sqrt{2})^2 + (4\sqrt{7})^2 = 210 \neq 308 = (2\sqrt{77})^2$; no, they cannot.

15. (a)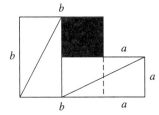

    (b) Since the square and double square are tiled by the same five shapes, their areas are equal. The respective areas are c^2 and $a^2 + b^2$, so $c^2 = a^2 + b^2$.

17.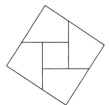

19. (a) By the Pythagorean theorem, $z^2 = a^2 + b^2$.

    (b) $a^2 + b^2 = c^2$ and $a^2 + b^2 = z^2$, so $z^2 = c^2$, and therefore $z = c$.

    (c) By the SSS property, $\triangle DEF \cong \triangle ABC$.

    (d) $\angle C$ is the angle corresponding to the right angle in the congruent triangle $\triangle DEF$.

20. (a)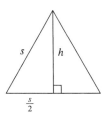

    $\left(\frac{s}{2}\right)^2 + (h)^2 = s^2$, by the Pythagorean theorem. Therefore, $h = \frac{\sqrt{3}}{2}s$.

22. (a) The scarecrow should say squares (not square roots), the two legs (not any two sides), and right triangle (not isosceles).

    (b) Suppose an isosceles triangle has sides of length x, x, and y. By the triangle inequality, $x + x > y$. Therefore, $2x > y$ and $\sqrt{2}\sqrt{x} > \sqrt{y}$, which means it is impossible for $\sqrt{x} + \sqrt{x}$ to equal $\sqrt{y}$. There is no isosceles triangle with Scarecrow's property.

25. Answers will vary. If the ladder cannot be vertical, the height is less than 24 ft. (The ladder can be vertical if it is bolted against a vertical wall.) If the base is 7 feet from the wall the top of the ladder is still nearly 23 feet off the ground.

27. Let d be the depth of the pond. The stem length is $d + 2$ (in feet). Held to the side, a right triangle is formed with legs of length 6 and d and hypotenuse of length $(d + 2)$. Then $6^2 + d^2 = (d + 2)^2$, or $36 + d^2 = d^2 + 4d + 4$, so $d = 8$.

29. (a) exact $= \sqrt{(200)^2 + (8000)(200)}$ miles $\doteq 1281$ miles;
    approximate $= \sqrt{(8000)(200)}$ miles $\doteq 1265$ miles

30. (a) $d \doteq 1.2\sqrt{100} = 12$ miles

31. The sum of the areas of the equilateral triangles on the legs equals the area of the equilateral triangle on the hypotenuse.

33. Let the original dimensions be l and w. The changed dimensions are $\frac{4}{3}l$ and $\frac{3}{4}w$. So area $= \left(\frac{4}{3}l\right)\left(\frac{3}{4}w\right) = lw$, the same as the original area.

35. Along large semicircle: $\frac{1}{2}(2 \cdot \pi \cdot 8 \text{ m}) = 8\pi$ m
    Along the two smaller semicircles:
    $\frac{1}{2}(2 \cdot \pi \cdot 3 \text{ m}) + \frac{1}{2}(2 \cdot \pi \cdot 5 \text{ m}) = 8\pi$ m
    The distances are the same.

Problem Set 12.4 (page 821)

1. (a) $S = 2 \cdot \frac{1}{2}(20 \text{ cm} + 15 \text{ cm})(12 \text{ cm}) + (2 \text{ cm})(13 \text{ cm} + 15 \text{ cm} + 12 \text{ cm} + 20 \text{ cm})$
   $= 540 \text{ cm}^2$
   $V = \frac{1}{2}(20 \text{ cm} + 15 \text{ cm})(12 \text{ cm})(2 \text{ cm}) = 420 \text{ cm}^3$

2. (a) The lateral surface area of the cone is $\frac{2\pi(6 \text{ in})}{2\pi(15 \text{ in})} = \frac{2}{5}$ the area of a circle of radius 15 in, so
   $S = \frac{2}{5}(\pi)(15 \text{ in})^2 + \pi(6 \text{ in})^2 = 126\pi \text{ in}^2 \doteq 396 \text{ in}^2$.
   Height of cone is $\sqrt{(15 \text{ in})^2 - (6 \text{ in})^2} = \sqrt{189}$ in, so $V = \frac{1}{3}\pi(6 \text{ in})^2(\sqrt{189} \text{ in}) \doteq 518 \text{ in}^3$.

3. (a) $S = 6 \cdot (20 \text{ mm})^2 - 2 \cdot \pi(2 \text{ mm})^2 + (20 \text{ mm})(2\pi)(2 \text{ mm}) \doteq 2626 \text{ mm}^2$
   $V = (20 \text{ mm})^3 - \pi(2 \text{ mm})^2(20 \text{ mm}) \doteq 7749 \text{ mm}^3$

4. (a) $S = 4\pi(4 \text{ ft})^2 + (20 \text{ ft})(2\pi)(4 \text{ ft}) = 224\pi \text{ ft}^2 \doteq 704 \text{ ft}^2$
   $V = \frac{4}{3}\pi(4 \text{ ft})^3 + \pi(4 \text{ ft})^2(20 \text{ ft}) \doteq 1273 \text{ ft}^3$

5. (a) Slant height $= \sqrt{(20 \text{ cm})^2 - (12 \text{ cm})^2} = 16$ cm

6. (a)

8. Let V_1 be the volume of the cylinder with height 8.5″ and radius $\frac{11}{2\pi}$″. Let V_2 be the volume of the cylinder with height 11″ and radius $\frac{8.5}{2\pi}$″.
   $V_1 = \pi\left(\frac{8.5}{2\pi}″\right)^2(11″) \doteq 63.24 \text{ in}^3$
   $V_2 = \pi\left(\frac{11″}{2\pi}\right)^2(8.5″) \doteq 81.85 \text{ in}^3$
   The shorter cylinder has the greater volume.
   $\left(\text{Note that } \frac{V_1}{V_2} = \frac{8.5}{11}.\right)$

10. (a) The circumference of the cone is
    $\frac{3}{4}2 \cdot \pi(4 \text{ in}) = 6\pi$ in, so the radius is
    $\frac{6\pi \text{ in}}{2\pi} = 3$ in.

11. Let s be the radius of the semicircle. Then the slant height of the cone is s. Let d be the diameter of the cone. Then the circumference of the cone is $\pi d = \frac{1}{2}2\pi s$, so $d = s$.

13. Let r be the radius of the sphere, so r is the radius of the cylinder and $2r$ is the height of the cylinder.
    area(sphere) $= 4\pi r^2$ square units;
    area(cylinder) $= 2 \cdot \pi r^2 + (2\pi r)(2r)$
    $= 6\pi r^2$ square units
    Thus, $\frac{\text{area(sphere)}}{\text{area(cylinder)}} = \frac{4\pi r^2}{6\pi r^2} = \frac{2}{3}$.

14. (a) $V = Bh = (7 \text{ cm})(4 \text{ cm})(3 \text{ cm})$
    $= 84 \text{ cm}^3$

15. (a) Since the diameter of the 16″ pizza is 2 times that of the 8″ pizza, the area of the 16″ pizza is $2^2 = 4$ times that of the 8″ pizza by the similarity principle. Therefore, it will feed 4 people.

(b) Since 1.4^2 is about 2.0, the area of one 14″ pizza is nearly 2 times the area of one 10″ pizza by the similarity principle. Hence, one 14″ pizza is nearly the same amount of pizza as two 10″ pizzas, but costs two dollars less. It is better to buy one 14″ pizza.

17. (a) Since each small circle has $\frac{1}{2}$ the diameter of the large circle, each small circle has $\frac{1}{4}$ the area. Thus the shaded region has $1 - \frac{1}{4} - \frac{1}{4} = \frac{1}{2}$ the area of the large circle.

19. (a) Doubling the radius increases the volume by a factor of 4, while halving the height halves the volume, so the volume is doubled. The new volume is 200 mL.

22. Suppose the area of the similar figure with straight side of length 1 is A. By the similarity principle, the areas of the figures erected on the sides of the triangle are then $a^2 A$, $b^2 A$, and $c^2 A$, where a, b, and c are the respective scale factors. Since $a^2 + b^2 = c^2$ (Pythagorean theorem), we get $a^2 A + b^2 A = c^2 A$, showing that the sum of the areas of the figures erected on the legs is equal to the area of the figure erected on the hypotenuse.

24. (a) Since $\frac{2}{16} = \frac{1}{8} = \left(\frac{1}{2}\right)^3$, the scale factor is $\frac{1}{2}$, and so the height is $\frac{1}{2}(6 \text{ in}) = 3$ in.

26. (a) By similarity, the height of the large complete pyramid is 8 cm. The frustum's volume is found by subtracting the volume of the removed pyramid from the volume of the complete pyramid.
$V = \frac{1}{3}(10 \text{ cm})^2(8 \text{ cm}) - \frac{1}{3}(5 \text{ cm})^2(4 \text{ cm})$
$= \frac{700}{3} \text{ cm}^3 \doteq 233 \text{ cm}^3$.

28. The volume of the ring is
$\pi \left(\frac{9}{16} \text{ in}\right)^2 \left(\frac{5}{4} \text{ in}\right) - \pi \left(\frac{1}{2} \text{ in}\right)^2 \left(\frac{5}{4} \text{ in}\right)$
$\doteq 0.26 \text{ in}^3$, so the ring weighs about
$\left(\frac{6 \text{ oz}}{1 \text{ in}^3}\right)(0.26 \text{ in}^3) = 1.56$ oz.

30. The volume of a box is
$(4 \text{ in})(5 \text{ in})(8 \text{ in}) = 160 \text{ in}^3$, so the volume of two boxes is 320 in^3. The volume of a "tub" is $\pi(3 \text{ in})^2(10 \text{ in}) \doteq 283 \text{ in}^3$, so two boxes is a better buy.

32. (a) The scale factor is $\frac{5}{13}$, so the weight is $\left(\frac{5}{13}\right)^3 (106.75) \doteq 6.07$ pounds.

34. (a) 6 ft = 72 in, so the scale factor is $\frac{72 \text{ in}}{6 \text{ in}} = 12$. Thus the volume would be increased by the factor $12^3 = 1728$

35. (a) Let s be the scale factor. Since the area of an A1 sheet is half that of an A0 sheet, $s^2 = \frac{1}{2}$, so $s = \frac{1}{\sqrt{2}}$.

(b) Using part (a), we see that the width x of an A0 sheet is $\frac{y}{\sqrt{2}}$ where y is the length. Then $xy = \frac{y^2}{\sqrt{2}} = 1 \text{ m}^2$, so
$y = 2^{1/4} \text{ m} \doteq 1.2$ m and
$x = \frac{1}{2^{1/4}} \text{ m} \doteq 0.84$ m.

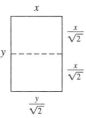

37. The diameter at the top is 3 times the diameter of the cylinder. Thus, the cross-sectional area is greater by a factor of $3^2 = 9$, so 9 inches of water in the cylinder correspond to 1 inch of rainfall.

38. **(b)** 3″ and $3 are not scale factors. The reasoning seems to suggest that a 6″ pizza should cost $2, and a 3″ pizza is given away with $1.

39. **(a)** $30^2 + 72^2 = 6084 = 78^2$; yes

41. $P = 3 + \sqrt{5} + 1 + \sqrt{5} + 4 + \sqrt{10} + \sqrt{2}$ 17 units;
    $A = \frac{1}{2}(2)(1) + \frac{1}{2}(2)(1) + (2)(4) + \frac{1}{2}(1)(1) + \frac{1}{2}(3)(1) = 12$ square units

42. $\frac{50 \text{ in}}{1 \text{ sec}} \cdot \frac{1 \text{ ft}}{12 \text{ in}} \cdot \frac{1 \text{ mi}}{5280 \text{ ft}} \cdot \frac{60 \text{ sec}}{1 \text{ min}} \cdot \frac{60 \text{ min}}{1 \text{ hr}} \doteq 2.84 \frac{\text{mi}}{\text{hr}}$

43. **(a)** $0.278 \text{ m} = 27.8 \times 10^{-2} \text{ m} = 27.8 \text{ cm}$

CLASSIC CONUNDRUM   **Strings and Balls (page 827)**

A string of length s will tightly encircle a sphere of radius r given by $r = \frac{s}{2\pi}$, since $s = 2\pi r$ is the circumference. A longer string of length S will make a circle of radius $R = \frac{S}{2\pi}$. Subtracting we see that $R - r = \frac{S-s}{2\pi}$. If S is 6′ larger than s, so $S - s = 6'$, then $R - r = \frac{6'}{2\pi} \doteq 1'$. That is, an extra 6′ of string will increase the radius of *any* circle by nearly a foot. The rabbit can easily run under the 6′ longer string that is everywhere 1′ above the equator.

Chapter Review Exercises (page 829)

1. **(a)** Centimeters

   (b) Millimeters

   (c) Kilometers

   (d) Meters

   (e) Hectares

   (f) Square kilometers

   (g) Milliliters

   (h) Liters

2. **(a)** 4 L

   (b) 190 cm

   (c) 200 m²

3. $(60 \text{ cm})(40 \text{ cm})(35 \text{ cm}) = 84{,}000 \text{ cm}^3$
   $= 84 \text{ L}$

4. $\frac{300 \text{ ft}}{3 \text{ sec}} \cdot \frac{1 \text{ mile}}{5280 \text{ ft}} \cdot \frac{60 \text{ sec}}{1 \text{ min}} \cdot \frac{60 \text{ min}}{1 \text{ hr}} \doteq 68 \frac{\text{miles}}{\text{hr}}$

5. Dissect the trapezoid by a horizontal line through M and rotate the bottom half by 180° as shown in the figure.

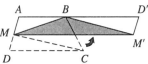

The triangle has half the area of the parallelogram $AD'M'M$ since they both have the same base and height, so it also has half the area of the trapezoid $ABCD$.

6. **(a)** 1 ft 4 in = 16 in, 4 ft = 48 in, so the area is (16 in)(48 in) = 768 in²
   $= 5\frac{1}{3} \text{ ft}^2$.

   (b) Dissect the figure with a diagonal from the bottom left vertex to the top right vertex to form two triangles. The area is
   $\frac{1}{2}(8 \text{ m})(9 \text{ m}) + \frac{1}{2}(3 \text{ m})(6 \text{ m}) = 45 \text{ m}^2$.

   (c) The area is
   $\frac{1}{2}(5 \text{ cm} + 7 \text{ cm})(3 \text{ cm}) = 18 \text{ cm}^2$.

7. **(a)** 11 square units

   (b) 9 square units

   (c) $9\frac{1}{2}$ square units

8. (a) $A = (3 \text{ ft})(4 \text{ ft}) + \frac{1}{2} \cdot \pi \cdot (1.5 \text{ ft})^2$
    $\doteq 15.5 \text{ ft}^2$;
    $P = 4 \text{ ft} + 3 \text{ ft} + 4 \text{ ft} + \frac{1}{2} \cdot 2\pi(1.5 \text{ ft})$
    $= 11 \text{ ft} + \pi(1.5 \text{ ft}) \doteq 15.7 \text{ ft}$

   (b) $A = \frac{3}{4} \cdot \pi(3 \text{ m})^2 = \frac{27}{4} \pi \text{ m}^2 \doteq 21.2 \text{ m}^2$;
    $P = 3 \text{ m} + 3 \text{ m} + \frac{3}{4} \cdot 2\pi(3 \text{ m})$
    $= 6 \text{ m} + \frac{9}{2} \pi \text{ m} \doteq 20.1 \text{ m}$

9. Solutions may vary. This is one possible solution. By the Pythagorean theorem, $x^2 = \left(\sqrt{6}\right)^2 + \left(\sqrt{30}\right)^2 = 36$, so $x = 6$. Next compute the area two ways.
   $A = \frac{1}{2}\left(\sqrt{6}\right)\left(\sqrt{30}\right) = 3\sqrt{5}$ and $A = \frac{1}{2}xy$,
   $\frac{1}{2}xy = 3\sqrt{5}$ or $xy = 6\sqrt{5}$. Since $x = 6$, $y = \sqrt{5}$.

10. The height of the cone is
    $\sqrt{(35 \text{ cm})^2 - (10 \text{ cm})^2} = \sqrt{1125} \text{ cm}$
    $\doteq 33.5 \text{ cm}$

11. $\sqrt{(4 \text{ in})^2 + (10 \text{ in})^2} = \sqrt{116} \text{ in}$;
    $\sqrt{(4 \text{ in})^2 + (12 \text{ in})^2} = \sqrt{160} \text{ in}$;
    $\sqrt{(10 \text{ in})^2 + (12 \text{ in})^2} = \sqrt{244} \text{ in}$;
    $\sqrt{(4 \text{ in})^2 + (10 \text{ in})^2 + (12 \text{ in})^2} = \sqrt{260} \text{ in}$

12. $P = 3 + \sqrt{2} + \sqrt{10} + \sqrt{5} + 1 + 5$
    $= 9 + \sqrt{2} + \sqrt{10} + \sqrt{5} \doteq 15.8$ units

13. (a) $V = \left[(10 \text{ ft})(20 \text{ ft}) + \frac{1}{2}(8 \text{ ft} + 20 \text{ ft})(8 \text{ ft})\right](30 \text{ ft}) = 9360 \text{ ft}^3$;
    $S = 2 \cdot \frac{1}{2}(8 \text{ ft} + 20 \text{ ft})(8 \text{ ft}) + 2 \cdot (10 \text{ ft})(20 \text{ ft}) + 2 \cdot (10 \text{ ft})(30 \text{ ft}) + 2 \cdot (10 \text{ ft})(30 \text{ ft}) + (8 \text{ ft})(30 \text{ ft}) + (20 \text{ ft})(30 \text{ ft})$
    $= 2664 \text{ ft}^2$

   (b) $V = \frac{1}{2} \cdot \frac{4}{3}\pi(7 \text{ m})^3 + \pi(7 \text{ m})^2(18 \text{ m}) = \left(1110\frac{2}{3}\right)\pi \text{ m}^3 \doteq 3489 \text{ m}^3$
    $S = \frac{1}{2} \cdot 4\pi(7 \text{ m})^2 + 2\pi(7 \text{ m})(18 \text{ m}) + \pi(7 \text{ m})^2 = 399\pi \text{ m}^2 \doteq 1253 \text{ m}^2$

   (c) $V = \frac{1}{3}\pi(5 \text{ cm})^2(8 \text{ cm}) + \frac{1}{2} \cdot \frac{4}{3}(5 \text{ cm})^3$
    $= 150\pi \text{ cm}^3 \doteq 471 \text{ cm}^3$
    $S = \frac{1}{2} \cdot 4\pi(5 \text{ cm})^2 + \frac{2\pi(5 \text{ cm})}{2\pi(8 \text{ cm})} \cdot \pi(8 \text{ cm})^2 = 90\pi \text{ cm}^2 \doteq 283 \text{ cm}^2$

14. $V(\text{sphere}) = \frac{4}{3}\pi(10 \text{ m})^3$ and $V(\text{four cubes}) = 4(10 \text{ m})^3$. Since $\pi > 3$, then $\frac{4}{3}\pi > 4$, showing that the sphere has the larger volume. Also, a sphere of radius 10 m can be inscribed in a cube with sides of length 20 m, which is the same as eight cubes with sides of length 10 m put together.

15. (a) The scale factor is $\frac{75 \text{ ft}}{50 \text{ ft}} = 1.5$: Heather needs $1.5(180 \text{ ft}) = 270$ ft of fencing.

    (b) Since Johan's garden area is $\left(\frac{1}{1.5}\right)^2$ times that of Heather's, he needs $\left(\frac{1}{1.5}\right)^2 45$ pounds = 20 pounds of fertilizer.

Chapter Test (page 831)

1. (a) Millimeters
   (b) Meters
   (c) Meters
   (d) Kilometers
   (e) Milliliters

(f) Liters

2. (a) 2161 mm = 2161×10^{-3} m
   = 216.1×10^{-2} m = 216.1 cm

   (b) 1.682 km = 1.682×10^3 m
   = $168,200 \times 10^{-2}$ m = 168,200 cm

   (c) 0.5 m^2 = 0.5(100 cm)^2 = 5000 cm

   (d) 1 ha = 10,000 m^2

   (e) 4719 mL = 4719×10^{-3} L = 4.719 L

   (f) 3.2 L = 3.2×1000 cm^3 = 3200 cm^3

3. (a) $(1147 \text{ in})\left(\frac{1 \text{ yd}}{36 \text{ in}}\right) \doteq 31.86$ yd

   (b) $(7942 \text{ ft})\left(\frac{1 \text{ mi}}{5280 \text{ ft}}\right) \doteq 1.5$ mi

   (c) $(32.4 \text{ yd}^2)\left(\frac{3 \text{ ft}}{1 \text{ yd}}\right)^2 = 291.6$ ft^2

   (d) $(9402 \text{ acres})\left(\frac{1 \text{ mi}^2}{640 \text{ acres}}\right) \doteq 14.69$ mi^2

   (e) $(7.6 \text{ yd}^3)\left(\frac{3 \text{ ft}}{1 \text{ yd}}\right)^3 = 205.2$ ft^3

   (f) $(5961 \text{ in}^3)\left(\frac{1 \text{ ft}}{12 \text{ in}}\right)^3 \doteq 3.45$ ft^3

4. $\frac{14 \text{ day}}{12 \text{ ft}} \cdot \frac{24 \text{ hours}}{1 \text{ day}} \cdot \frac{60 \text{ min}}{1 \text{ hour}} \cdot \frac{1 \text{ ft}}{12 \text{ in}}$
   = 140 $\frac{\text{min}}{\text{in}}$,
   so it takes 140 minutes to grow one inch.

5.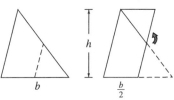

   area(triangle) = area(parallelogram)
   = $\left(\frac{b}{2}\right)(h) = \frac{1}{2}bh$

6. (a) 8 cm^2

   (b) 8 cm^2

   (c) 8.5 cm^2

7. $A = \frac{1}{2} \cdot (6 \text{ cm})(16 \text{ cm}) + \frac{1}{2}(15 \text{ cm})(16 \text{ cm})$
   = 168 cm^2
   Sides are 10 cm and 17 cm by the Pythagorean theorem, so P = 54 cm.

8. $A = \frac{1}{2}(9 \text{ ft})(12 \text{ ft}) + (8 \text{ ft})(12 \text{ ft}) - \frac{1}{2}\pi(4 \text{ ft})^2$
   $\doteq$ 125 ft^2
   $P = 15 \text{ ft} + 9 \text{ ft} + 8 \text{ ft} + 12 \text{ ft} + \frac{1}{2} \cdot 2\pi(4 \text{ ft})$
   $\doteq$ 56.6 ft

9. (a) $\frac{2}{3}\pi(5 \text{ in})^2 \doteq 52.4$ in^2

   (b) $\frac{1}{2}(2.6 \text{ m} + 1.4 \text{ m})(3 \text{ m}) = 6.0$ m^2

   (c) $\frac{1}{12} \cdot \pi(24 \text{ cm})^2 \doteq 151$ cm^2

10. Let the radius of the circle be r. Then the circumscribed square has sides of length $2r$ and the inscribed square has sides of length $\sqrt{2}r$. $\frac{\text{area(inscribed)}}{\text{area(circumscribed)}} = \frac{\left(\sqrt{2}r\right)^2}{(2r)^2} = \frac{2r^2}{4r^2} = \frac{1}{2}$

    Alternate solution: The scale factor of the large to the small square is $\frac{1}{\sqrt{2}}$, so the small square has $\left(\frac{1}{\sqrt{2}}\right)^2 = \frac{1}{2}$ the area of the large square.

11. Cross-sectional view:

    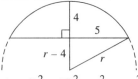

    $(r-4)^2 + (5)^2 = r^2$, or $r^2 - 8r + 16 + 25 = r^2$. Then $r = \frac{16+25}{8 \text{ mm}} = 5.125$ mm.

12. By the Pythagorean theorem, the ladder reaches $\sqrt{(15 \text{ ft})^2 - (6 \text{ ft})^2} = \sqrt{189}$ ft $\doteq 13.7$ ft.

13. (a) $P = \sqrt{52} + \sqrt{13} + \sqrt{65} \doteq 18.9$ units

    (b) Yes, it is a right triangle since $\left(\sqrt{52}\right)^2 + \left(\sqrt{13}\right)^2 = 65 = \left(\sqrt{65}\right)^2$.

14. (a) The hypotenuse of the triangle has length $\sqrt{(7 \text{ m})^2 + (24 \text{ m})^2} = 25$ m.
    $S = 2 \cdot \frac{1}{2}(7 \text{ m})(24 \text{ m}) + (7 \text{ m} + 24 \text{ m} + 25 \text{ m})(5 \text{ m}) = 448 \text{ m}^2$;
    $V = \frac{1}{2}(7 \text{ m})(24 \text{ m})(5 \text{ m}) = 420 \text{ m}^3$

    (b) $S = 2 \cdot \frac{1}{2}\pi(4 \text{ in})^2 + \frac{1}{2} \cdot 2\pi(4 \text{ in})(6 \text{ in}) + (6 \text{ in})(8 \text{ in}) \doteq 173.7 \text{ in}^2$;
    $V = \frac{1}{2}\pi(4 \text{ in})^2(6 \text{ in}) \doteq 150.8 \text{ in}^3$

    (c) The slant height is $\sqrt{(8 \text{ ft})^2 + (6 \text{ ft})^2} = 10$ ft.
    $S = 4 \cdot \frac{1}{2}(12 \text{ ft})(10 \text{ ft}) + (12 \text{ ft})^2 = 384 \text{ ft}^2$;
    $V = \frac{1}{3}(12 \text{ ft})^2(8 \text{ ft}) = 384 \text{ ft}^3$

    (d) The slant height is $\sqrt{(12 \text{ cm})^2 + (5 \text{ cm})^2} = 13$ cm.
    $S = \pi(5 \text{ cm})^2 + \frac{2\pi(5 \text{ cm})}{2\pi(13 \text{ cm})} \cdot \pi(13 \text{ cm})^2 = 90\pi \text{ cm}^2 \doteq 283 \text{ cm}^2$
    $V = \frac{1}{3}\pi(5 \text{ cm})^2(12 \text{ cm}) = 100\pi \text{ cm}^3 \doteq 314 \text{ cm}^3$

15. (a) $S = 4\pi(10 \text{ m})^2 = 400\pi \text{ m}^2 \doteq 1257 \text{ m}^2$
    $V = \frac{4}{3}\pi(10 \text{ m})^3 \doteq 4189 \text{ m}^3$

    (b) $S = \pi(5 \text{ cm})^2 + 2\pi(5 \text{ cm})(6 \text{ cm}) + \frac{1}{2} \cdot 4\pi(5 \text{ cm})^2 = 135\pi \text{ cm}^2 \doteq 424 \text{ cm}^2$
    $V = \pi(5 \text{ cm})^2(6 \text{ m}) + \frac{1}{2} \cdot \frac{4}{3}(5 \text{ cm})^3 \doteq 733 \text{ cm}^3$

16. The scale factor of Papa Bear to Mama Bear is $\frac{4}{5}$ and the scale factor of Mama Bear to Baby Bear is $\frac{1}{2}$.

	Papa Bear	Mama Bear	Baby Bear
Length of Suspenders	50	40	20 in
Weight	468.75	240	30 lb
Number of fleas	6000	3840	960

17. $V(\text{peel}) = \frac{4}{3}\pi(2.5 \text{ m})^3 - \frac{4}{3}\pi(1.75 \text{ in})^3 \doteq 43.0 \text{ in}^3$, and $V(\text{grapefruit}) = \frac{4}{3}\pi(2.5 \text{ in})^3 \doteq 65.4 \text{ in}^3$.
    About 66% is peel. Alternate solution:
    The scale factor is $\frac{2.5 - 0.75}{2.5} = 0.7$.
    Since $(0.7)^3 = 0.343$, it follows that 34.3% of the grapefruit is not peel, and 65.7% is peel.

Chapter 13

JUST FOR FUN   Quadrilateral + Quadrilateral = Parallelogram? (page 838)

Since the sum of the interior angles of a quadrilateral is 360°, we might start by guessing that the four angles corresponding to the intact corners of the quadrilateral should go in the center of our new figure. Arranging the pieces so that congruent sides line up, we find that this strategy works.

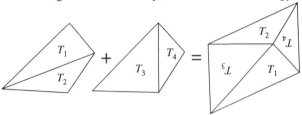

Problem Set 13.1 (page 855)

1. (a) Not a rigid motion; distances between particular cards will change.

   (c) No, distances between particular pieces almost certainly will have changed.

2. (a) The figure is moved 3 units to the right.

4. (a) $360° - 60° = 300°$

   (c) $43°$, since the remainder after dividing 3643 by 360 is 43.

5. (a)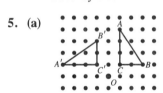

6. (a) The center of rotation is the same distance from A as from A', so it is on the perpendicular bisector of $\overline{AA'}$. (Similarly, it is on the perpendicular bisector of $\overline{BB'}$, but this turns out to be the same line.) By using trial and error among the points on the perpendicular bisector, we find that the only point that can produce the desired transformation of both points simultaneously is point O, shown below.

   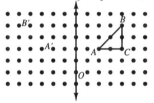

8. (a) O is the midpoint of segment $\overline{PP'}$.

   (b) Draw the line $\overleftrightarrow{AO}$. Set the compass to distance AO and mark off this distance from O on $\overleftrightarrow{AO}$ away from A. This is A'. Likewise for points B and C.

9. Since the transformation preserves orientation and involves a net rotation of 360° (which is equivalent to 0°), it can only be a translation. A little experimentation shows that the translation is in the direction of ray $\overrightarrow{O_1O_2}$ and its distance is $2x$, where x is the distance from O_1 to O_2.

   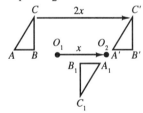

172   *Chapter 13: Mathematical Reasoning for Elementary Teachers*

11. (a) Draw the perpendicular bisector of $\overline{PP'}$. Since $\overline{PP'}$ is horizontal, the line of reflection is vertical.

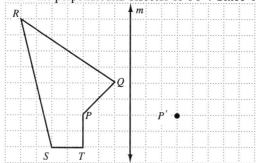

13. (a), (b) For part (a), notice that the slide arrow is 7 units long, so A, B, C, D, E is 7 units to the right of *ABCDE*. For part (b), reflect A, B, C, D, E, across the line of reflection to find the image of *ABCDE* under the glide reflection.

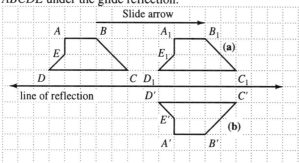

15. (a)

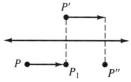

16. (a) The distance from m_1 to m_2 must be 3 units, or half the length of the slide arrow.

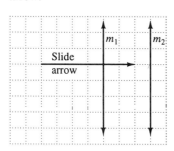

18. (c) $\frac{1}{2} \cdot 180° = 90°$

(e) $\frac{1}{2} \cdot 90° = 45°$

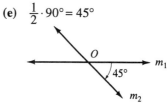

19. (a) Since $A'P' = AP = 2$ cm, P' is some point on the circle of radius 2 cm centered at A'.

20. (a) If the lines are parallel, then all possible segments $\overline{PP'}$ must be parallel. This is true for a reflection or a translation. But a reflection would only give one perpendicular bisector (instead of parallel lines), so the basic motion is a translation.

21. Since each of the turns preserves orientation, the combined transformation must preserve orientation, so it is either a translation or a rotation. But there is an overall turn of 180°, so it remains only to determine the center of this rotation. On a 1 cm square grid, the two 90° rotations take O_1 to O_1' (and O_2 to O_2') as shown. Since the rotation center of a half turn is half way between any point and its image, it must be the point O shown below. The motion is equivalent to a 180° rotation (half turn) about the point O.

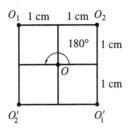

23. (a), (b) Reflection of $\triangle ABC$ across m_1 gives $\triangle A_1B_1C_1$. Reflection of $\triangle A_1B_1C_1$ across m_2 gives $\triangle A_2B_2C_2$. Finally, reflection $\triangle A_2B_2C_2$ across m_3 gives $\triangle A'B'C'$. For part (b), l is the line that contains the midpoints of $\overline{AA'}$, $\overline{BB'}$, and $\overline{CC'}$. (It is also the perpendicular bisector of any one of these.)

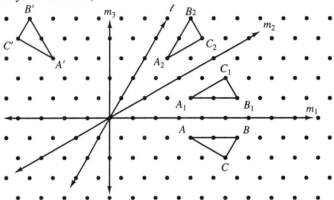

25. (a), (b) m_1 is the perpendicular bisector of $\overline{AA'}$. m_2 is the perpendicular bisector of $\overline{BB'}$. The image of C_1 across m_2 is C'.

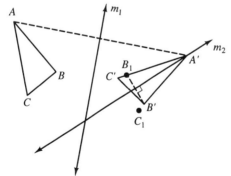

(c) The rigid motion that takes $\triangle ABC$ to $\triangle A'B'C'$ is a reflection across m_1 followed by a reflection across m_2. Since two consecutive reflections across intersecting lines are equivalent to a rotation, the basic rigid motion is a rotation about the point P of intersection of m_1 and m_2, through an angle twice the measure x of the directed angle from line m_1 toward line m_2.

27. (a) A translation: Six reflections give an orientation preserving rigid motion, so it is either a rotation or a translation. Since a rotation has a fixed point (namely the rotation center), the motion is a translation.

28. Translate $ABCD$ to $A'B'C'D'$ using the slide arrow that takes A to C. Then BB' and DD' are parallel, as are BD and BD', so $BB'D'D$ is a parallelogram. This parallelogram can be subdivided into $\triangle B'CB$, $\triangle BCD$, $\triangle DCD'$, and $\triangle D'CB'$, which are congruent to $\triangle ABC$, $\triangle BCD$, $\triangle CDA$, and $\triangle DAB$, respectively.

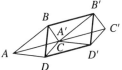

30. (a) Since A and B have opposite orientations (which is most easily observed by looking at the tail) and are not reflections of each other, the motion is a glide reflection.

31. The mirror only needs to be 31″ tall, half of Estelle's height of 62″. The top of the mirror can be 3″ lower than the top of Estelle's head; that is, the top is 59″ above the floor. It does not matter how far away Estelle stands; her reflection is always twice as far from her as the mirror.

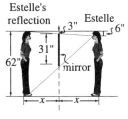

32. (a) $\overline{RS'}$ is the reflection of $\overline{RS}$ across m, so $RS' = RS$ since a reflection preserves all distances. Similarly, $QS' = QS$.

34. A single mirror reverses orientation, so the double reflection seen in a corner mirror preserves orientation. The corner mirror reflection of your right hand will appear as a right hand.

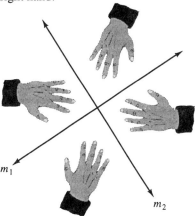

36. To see that the octagon tiles the plane, see the quadrilateral tilings in Example 10.14, page 656.

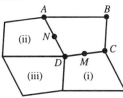

38. The turtle needs to draw the triangle and then retrace his steps along the last side. The next triangle is drawn continuing from this position. Since there are five triangles, this is repeated five times.
```
to spin.scalene
  repeat 5 [scalene lt 55.77 fd 60]
end
```

40. (a) Note that the interior angles on the outer boundary of the figure can only be created using hexagons. Therefore, the figure consists of 6 hexagons, so `:nsides = 6` and `:nreps = 6`.

41. (a) This procedure draws the original parallelogram, moves the turtle 50 steps up (with the pen lifted), and then draws the translated parallelogram.

```
            to translate
                parallelogram.2
                pu fd 50 pd
                parallelogram.2
            end
```

43. $2x + 3x + 4x = 180$, so $x = 20$, giving angles of measure $40°$, $60°$, and $80°$.

45. Any rhombus that is not a square can be used to form a counterexample. In the example shown, the midpoints form the non-square rectangle *KLMN*.

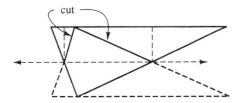

JUST FOR FUN   Mirror Magic (page 871)

Reflecting the mirror across the line through the midpoints of the shorter sides determines where Amy can make two straight cuts. The resulting pieces are a kite and two isosceles triangles. Each of these pieces has a line of symmetry, and so the pieces can be turned over and glued in place.

JUST FOR FUN   The Penny Game (page 873)

Lynn can always place a penny at the position that is the image of Kelly's last move under a half-turn centered at the center of the table. Since Lynn can always make a move, Kelly will be the first player unable to find space for an additional penny on the table.

Problem Set 13.2 (page 875)

1. (a) 6 lines of symmetry

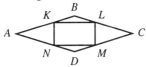

(c) No lines of symmetry

3. (a) Answers will vary. One possibility is an isosceles triangle.

4. (a)

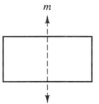

(c)

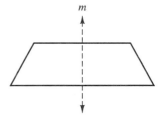

7. (a) Unfolding the horizontal fold produces the figure on the left, below. Unfolding the vertical fold then produces the figure on the right.

8. (a) One vertical line of symmetry

9. Note that some of the patterns shown represent light reflections rather than actual patterns in the wheel covers themselves. These should be ignored when determining symmetry.

(a) Five lines of symmetry and $72°$ rotation symmetry

(b) $72°$ rotation symmetry

11. (a)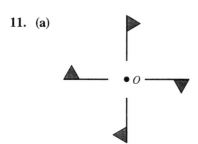

12. Note that any regular n-gon has n lines of symmetry. If n is odd, there is a line of symmetry passing through each vertex and the midpoint of the opposite side. If n is even, there are $\frac{n}{2}$ lines of symmetry through pairs of opposite vertices, and $\frac{n}{2}$ lines of symmetry through the midpoints of opposite sides, for a total of n lines of symmetry.

    (a) Equilateral triangle

    (b) Square

13. (a) 0, I, 8

    (b) 0, I, 3, 8

    (c) 0, I, 8

    (d) 0, I, 8

16. (a) Translation symmetry and vertical line symmetry: $m1$

17. (a) Translation symmetry, vertical line symmetry, horizontal line symmetry, glide reflection symmetry, and half-turn symmetry: mm

18. (a) Translation symmetry; vertical line symmetry (and glide symmetry) across lines through a column of As or midway between columns of As.

    (c) Translation symmetry; 180° rotation symmetry about a point which is the center of N, or midway between the centers of any two Ns along a horizontal, vertical, or diagonal line. Note that there is no reflection or glide symmetry because of orientation.

    (e) Translation symmetry; vertical line symmetry (and glide symmetry) across lines that are midway between columns; horizontal glide symmetry across lines that are midway between rows; 180° rotation symmetry about points that are midway between an adjacent p and d or an adjacent b and q.

19. (a) No letter is the same as another letter written backwards. That is, no letter or digit reflects vertically into a different letter or digit, so it must reflect into itself.

21. The pattern would be the same after reflecting across the vertical line and rotating 180°, and this pair of transformations is equivalent to a reflection across a horizontal line. Therefore, the pattern must also have a horizontal line of symmetry, so it would be an mm pattern.

24. (a) The pattern has no vertical line of symmetry and, disregarding the color scheme, it has horizontal glide symmetry. Its type is $1g$.

    (c) The pattern has vertical line symmetry and horizontal glide symmetry. Its type is mg.

26. (a) The motion is a translation. The pattern does not have vertical line symmetry, horizontal glide reflection, or half-turn symmetry. Its type is 11.

| p | p | p | p | p | p |

    (c) The motion is a half-turn. The pattern has half-turn symmetry only. Its type is 12.

| p | d | p | d | p | d |

28. (a) Three directions of reflection symmetry (including vertical); the same three directions of glide symmetry; 120° rotation symmetry

29. (a) 3, parallel to the faces of the prism

30. (a) As left-handed people know well, not all scissors are symmetric.

    (c) A dress shirt is not symmetric due to buttons and pockets.

    (e) Tennis rackets have two planes of bilateral symmetry (ignoring such details as how they are strung and wrapping around the handle).

31. (a) There is symmetry across the main diagonal line of entries in the sense that the same numbers appear on both sides of this diagonal. In other words, addition is commutative.

35. (a) Let x be the measure of each interior reflex angle. Then $4x + 8 \cdot 120° = (12 - 2) \cdot 180°$, so $x = 210°$. This means that the turtle must turn left $210° - 30°$ after each time it completes the `side` procedure.

```
to fig4
    repeat 4 [ side lt 30 ]
end
```

36. (a) Yes, it is a rotation.

(b) No, the distance between opposite corners has changed.

37. (a) $130° + 220° = 350°$

38. (a) Note that $A = A_1 = A_2$ and $C = C_1$.

(b) Glide reflection, 3 units right and reflect across the line l parallel to $\overline{AB}$ and midway between C and $\overline{AB}$.

Problem Set 13.3 (page 887)

1. (a) A' is the point on $\overrightarrow{OA}$ such that $OA' = 2 \cdot OA$.

B' is the point on $\overrightarrow{OB}$ such that $OB' = 2 \cdot OB$.

C' is the point on $\overrightarrow{OC}$ such that $OC' = 2 \cdot OC$.

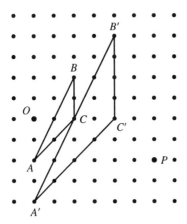

(b) A' is the point on $\overrightarrow{PA}$ such that $PA' = \frac{1}{2} \cdot PA$.

B' is the point on $\overrightarrow{PB}$ such that $PB' = \frac{1}{2} \cdot PB$.

C' is the point on $\overrightarrow{PC}$ such that $PC' = \frac{1}{2} \cdot PC$.

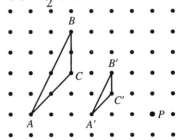

3. (a) The smallest square has scale factor $\frac{1}{3}$. The next smallest square has scale factor $\frac{2}{3}$. The largest square has scale factor $\frac{4}{3}$.

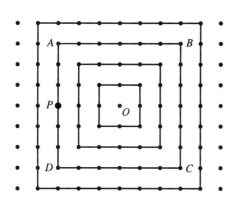

4. (a) The scale factor is $\dfrac{A'B'}{AB} = \dfrac{28}{21} = \dfrac{4}{3}$.

(b) Find the intersection of two lines such as $\overleftrightarrow{AA'}$ and $\overleftrightarrow{BB'}$. Alternately, find the point on $\overrightarrow{A'A}$ (for example) that is 4 times as far from A' as A is, implying $OA' = \dfrac{4}{3} \cdot OA$.

(c) Since $\dfrac{4}{3} \cdot BC = B'C' = 24,\ BC = \dfrac{3}{4} \cdot 24 = 18$.
$m(\angle D) = m(\angle D') = 80°$

6. (a) The center is at P, the intersection of $\overleftrightarrow{DG}$ and $\overleftrightarrow{EH}$. The scale factor is $\dfrac{GI}{DF} = \dfrac{6}{4} = \dfrac{3}{2}$.

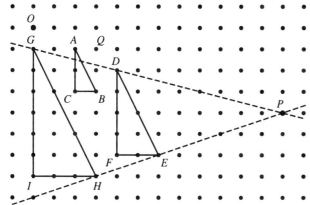

7. $\triangle A_1B_1C_1$ is obtained by applying a half-turn about H to $\triangle ABC$. $\triangle A'B'C'$ is obtained by applying a size transformation centered at point O with scale factor 2 to $\triangle A_1B_1C_1$.

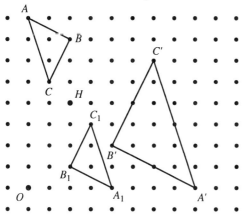

9. We need to rotate the figure 90° counterclockwise so that, for example, $\overline{A_1B_1}$ will be parallel to $\overline{A'B'}$. Then we can perform a size transformation whose center is found by intersecting two lines such as $\overleftrightarrow{B_1B'}$ and $\overleftrightarrow{D_1D'}$. Answers will vary; one possibility is the following.
Rotate 90° counterclockwise about B. Then perform a size transformation centered at P with scale factor 2.

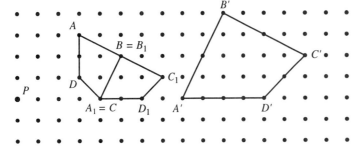

11. No; a quadrilateral with sides of those lengths can be in many shapes with different angles.

16. A size transformation about A with scale factor $\frac{AP}{AQ}$ will yield a cycloid passing through P. So the diameter of the circle would be $\frac{AP}{AQ} \cdot 1 = \frac{AP}{AQ}$.

17. The line $\overleftrightarrow{PP'}$ passes through O, so O is the intersection of $\overleftrightarrow{PP'}$ and line l. Since $\overline{PQ}$ and $\overline{P'Q'}$ are parallel, Q' lies on the line through P' that is parallel to $\overline{PQ}$. Since Q' also lies on $\overrightarrow{OQ}$ (which is part of line l), Q' is the intersection of line l and the line through P' that is parallel to $\overline{PQ}$.

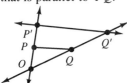

19. (a) Since A is not moved, the center is A. The scale factor is $\frac{AF}{AB} = \frac{5}{1} = 5$.

21. The larger grid is a size transformation that can easily be drawn. The short curves within each square can be drawn with good accuracy.

23. (a) All parts of the rubber band stretch by the same factor. Also, the band stretches to form a straight line segment.

    (b) Since $OP' = 2 \cdot OP$, the scale factor is 2.

(c) Since we desire $OP' = 3 \cdot OP$, the knot should be $\frac{1}{3}$ of the way from the pinned point to the other end of the loop.

27.

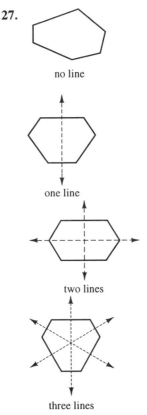

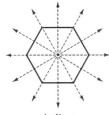

six lines

Hexagons with four, five, and seven or more lines of symmetry are not possible.

29. One way to describe the inverse motion is simply to reverse each step, and do them in the opposite order: Translate left by 6 feet, reflect across l, and then do a half-turn about P. (Other ways are possible.)

CLASSIC CONUNDRUM   Inverting the Tetractys (page 891)

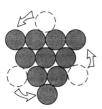

Chapter 13 Review Exercises (page 893)

1. Move each point 3 units right and 1 unit down.

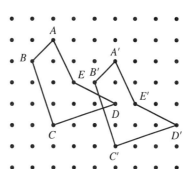

2. Find the perpendicular bisectors of $\overline{AA'}$ and $\overline{BB'}$. Their intersection is the turn center O, and the measure of $\angle AOA'$ is the turn angle, approximately 63° clockwise or −63°.

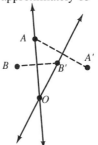

3. Reflection across line l. The line l is determined by any two of the midpoints of $\overline{AA'}$, $\overline{BB'}$, and $\overline{CC'}$.
It is also the perpendicular bisector of each of these segments.

4. Draw any two vertical lines two inches apart. There are then three ways to choose the order of the successive lines of reflection.

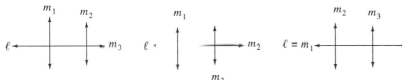

5. (a) 1, about 20° above horizontal

   (b) 2, horizontal and vertical

   (c) None, due to pattern interweaving

   (d) None, due to pattern interweaving

   (e) 3, including vertical

   (f) Infinitely many (all lines through the center point, since the figure has circular symmetry)

6. (a) 0° (no rotation symmetry)

(b) 0°, 180°

(c) 0°, 180°

(d) 0°, 72°, 144°, 216°, 288°

(e) 0°, 120°, 240°

(f) Any angle

7. (a) Translation symmetry and vertical line symmetry (type $m1$)

   (b) Translation symmetry and horizontal line symmetry (type $1m$)

8. O is a point on $\overleftrightarrow{AA'}$ for which $OA' = \frac{3}{2} \cdot OA$, implying that $OA = 2 \cdot AA'$, as shown.

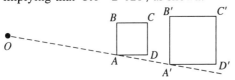

9. Since F and F' are similar figures with scale factor $k \cdot \left(\frac{1}{k}\right) = 1$, F and F' are congruent.

   Thus some rigid motion will transform F to F'. Since no rotation or reflection takes place, this rigid motion must be a translation.

10. By the Pythagorean theorem, each diagonal of $ABCD$ has measure $\sqrt{2}AB$. Each diagonal of $JKLM$ has measure AB, so a size transformation about the center P of $ABCD$ with scale factor $\frac{1}{\sqrt{2}} = \frac{\sqrt{2}}{2}$ should be accomplished, along with a 45° rotation about P. These transformations can be done in either order.

Chapter 13 Test (page 894)

1. Move C 3 units right and 2 units down to find C' (2 units right of B').
   Move B' 3 units left and 2 units up to find B (2 units left of C).

2. (a) The center is P, the intersection of the perpendicular bisectors of $\overline{AA'}$ and $\overline{BB'}$. The rotation angle is 90°.

(b)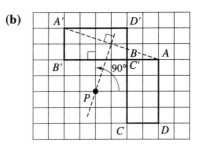

3. Construct the perpendicular bisector of AA'. This is the line of reflection. Label B and C, the points of intersection of the line of reflection with the rays of $\angle A$. Since these points are unchanged by the reflection, the desired angle is $\angle BA'C$.

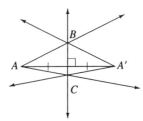

4. The line of reflection contains the midpoints of $\overline{AA'}$, $\overline{BB'}$, and so on.

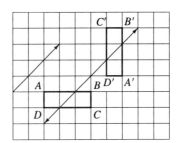

5. Various pairs of lines are possible. The distance between the lines must be half the distance between P and P'. In addition, the lines must be perpendicular to $\overline{PP'}$ and m_2 must be "above" m_1.

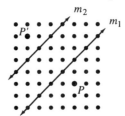

6. Many pairs of lines are possible. The two lines need to intersect at point O and the directed angle between the lines should be half the

measure of ∠QOQ'.

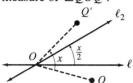

7. (a) Glide reflection, since the first two reflections accomplish a translation in a direction parallel to the third reflection.

   (b) Reflection, since the 3 lines of reflection are parallel.

   (c) Reflection, since the 3 lines of reflection are concurrent.

   (d) Glide reflection, since the 3 lines of reflection are neither parallel nor concurrent.

8. (a) Rotation (half-turn)

   (b) Rotation

9. (a)

   (b)

   (c)

10. (a) 180°

    (b) 360° (only the identity rotation)

    (c) $\frac{360°}{7}$

11. (a) Point symmetry, and two diagonal lines of symmetry.

    (b) Four lines of symmetry, and 90° rotation symmetry.

12. (a) Translation symmetry and glide reflection symmetry (type 1g)

    (b) Translation symmetry, vertical line symmetry, horizontal reflection (and glide reflection) symmetry, and half-turn symmetry (type mm)

    (c) Translation symmetry and half-turn symmetry (type 12)

    (d) Translation symmetry, vertical line symmetry, horizontal glide reflection symmetry, and half-turn symmetry (type mg)

13. A rotation symmetry of a wallpaper pattern can have size 60°, 90°, 120°, or 180°, but no other size is possible.

14. (a) The rotation symmetry is 90°.

    (b) The rotation symmetry is 120°.

15. (a) $k = \frac{DE}{AB} = \frac{16}{8} = 2$

    (b) $EF = 2 \cdot BC = 2 \cdot 12 = 24$

    (c) $AC = \frac{1}{2} \cdot DF = \frac{1}{2} \cdot 36 = 18$

16. (a) Since A is unmoved, the center is A.

    (b) The scale factor is $\frac{AV}{AB} = \frac{8}{8+4} = \frac{2}{3}$.

    (c) $VW = \frac{2}{3} \cdot 20 = \frac{40}{3}$

    (d) $ZF = \frac{3}{2} \cdot 18 - 18 = 27 - 18 = 9$

17. Many transformations are possible. Here is one sequence: translate the square so A is taken to A'; rotate about A' by 45°; perform a size transformation about A' with scale factor $\frac{3\sqrt{2}}{2}$ (since $A'B' = 3\sqrt{2}$ and $AB = 2$).

Chapter 14

Problem Set 14.1 (page 911)

1. **(a)** The second coordinate is zero.

   (c) The first coordinate is positive and the second coordinate is negative.

2. (a), (c), (e), (g), (i)

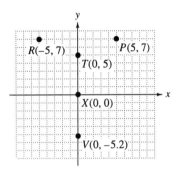

6. **(a)** $M(r, s) = M\left(\frac{2+6}{2}, \frac{7+1}{2}\right) = M(4, 4)$

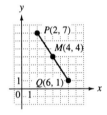

(c) $M(r, s) = M\left(\frac{0+(-4)}{2}, \frac{7+1}{2}\right)$
$= M(-2, 4)$

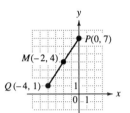

(e) $M(r, s) = M\left(\frac{2+7}{2}, \frac{3+(-3)}{2}\right)$
$= M\left(4\frac{1}{2}, 0\right)$

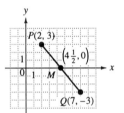

7. Use the theorem on page 907, where $a = 1$, $b = -7$, $c = -7$, and $d = 9$.

   (a) Let $t = \frac{1}{4}$. $((1-t)a + tc, (1-t)b + td)$
   $= \left(\frac{3}{4} \cdot 1 + \frac{1}{4} \cdot (-7), \frac{3}{4} \cdot (-7) + \frac{1}{4} \cdot 9\right)$
   $= (-1, -3)$

   (c) Let $t = \frac{5}{4}$. $((1-t)a + tc, (1-t)b + td)$
   $= \left(-\frac{1}{4} \cdot 1 + \frac{5}{4} \cdot (-7), -\frac{1}{4} \cdot (-7) + \frac{5}{4} \cdot 9\right)$
   $= (-9, 13)$

(d)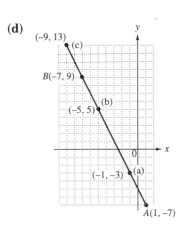

9. (a) $\sqrt{(4-(-2))^2+(13-5)^2} = \sqrt{6^2+8^2} = \sqrt{36+64} = \sqrt{100} = 10$

(c) $\sqrt{(8-0)^2+(-8-7)^2} = \sqrt{8^2+(-15)^2} = \sqrt{64+225} = \sqrt{289} = 17$

10. (a) $(RS)^2 = \left(\sqrt{(7-1)^2+(10-2)^2}\right)^2 = \left(\sqrt{6^2+8^2}\right)^2 = \left(\sqrt{36+64}\right)^2 = \left(\sqrt{100}\right)^2 = 100$

$(RT)^2 = \left(\sqrt{(5-1)^2+(-1-2)^2}\right)^2 = \left(\sqrt{4^2+(-3)^2}\right)^2 = \left(\sqrt{16+9}\right)^2 = \left(\sqrt{25}\right)^2 = 25$

$(ST)^2 = \left(\sqrt{(5-7)^2+(-1-10)^2}\right)^2 = \sqrt{(-2)^2+(-11)^2} = \left(\sqrt{4+121}\right)^2 = \left(\sqrt{125}\right)^2 = 125$

Since $(RS)^2 + (RT)^2 = 100 + 25 = 125 = (ST)^2$, by Pythagorean theorem $\triangle RST$ is a right triangle.

(c)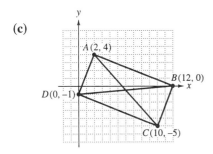

12. (a) Label the points $A(1, 1)$, $B(1, -4)$, and $C(6, -4)$. Then

$AB = \sqrt{(1-1)^2+(-4-1)^2} = \sqrt{0+25} = \sqrt{25} = 5$

$BC = \sqrt{(6-1)^2+(-4-(-4))^2} = \sqrt{25+0} = \sqrt{25} = 5$

$AC = \sqrt{(6-1)^2+(-4-1)^2} = \sqrt{25+25} = \sqrt{50} = 5\sqrt{2}$

Since $(AB)^2 + (BC)^2 = (AC)^2$ and $AB = BC$, the triangle is right and isosceles.

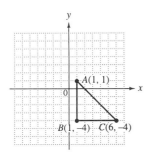

(c) Label the points $A(-1, 3)$, $B(2, 5)$, and $C(7, -3)$. Then
$$AB = \sqrt{(2-(-1))^2 + (5-3)^2} = \sqrt{9+4} = \sqrt{13}$$
$$BC = \sqrt{(7-2)^2 + (-3-5)^2} = \sqrt{25+64} = \sqrt{89}$$
$$AC = \sqrt{(7-(-1))^2 + (-3-3)^2} = \sqrt{64+36} = \sqrt{100} = 10$$
Since the square of each side length is less than the sum of the squares of the other two side lengths, and no two side lengths are equal, the triangle is acute and scalene.

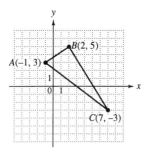

13. (a)

a	b	P(a, b)
−5	11	(−5, 11)
−4	10	(−4, 10)
−3	9	(−3, 9)
−2	8	(−2, 8)
−1	7	(−1, 7)
0	6	(0, 6)
1	5	(1, 5)
2	4	(2, 4)
3	3	(3, 3)
4	2	(4, 2)
5	1	(5, 1)

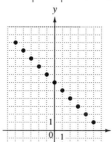

15. (a) Plot the points (−6, 5), (−5, 5), (−4, 5), (−3, 5), (−2, 5), (−1, 5), (0, 5), (1, 5), (2, 5), (3, 5), (4, 5), (5, 5), and (6, 5).

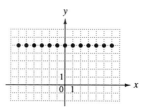

21. In each case, Q needs to have the same y-coordinate as P, and the x-coordinate of Q needs to be the additive inverse of the x-coordinate of P.

(a)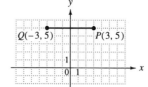

(c)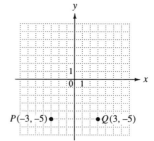

(e)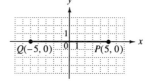

186 *Chapter 14: Mathematical Reasoning for Elementary Teachers*

22. In each case, T needs to have the same x-coordinate as S, and the y-coordinate of T needs to be the additive inverse of the y-coordinate of S.

(a)

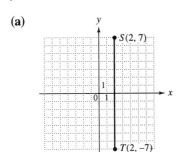

(c)

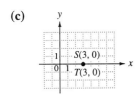

(e)

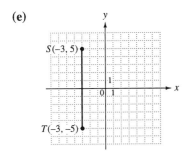

23. (a)

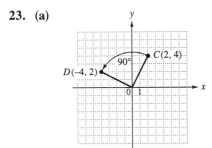

(c)

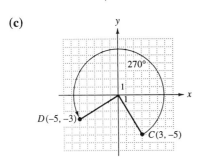

25. In each case, the x- and y-coordinates of H must be the additive inverses of the corresponding coordinates of G.

(a)

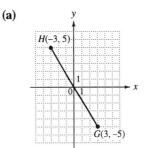

(c)

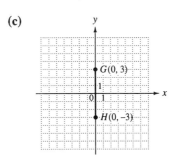

(e)

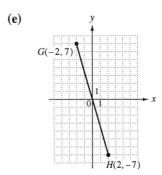

26. (a) Note that B can be obtained by sliding A to the right 3 and up 5. Therefore, C can be obtained by sliding D to the right 3 and up 5. The coordinates of C are $(7 + 3, 0 + 5) = (10, 5)$. Hence $r = 10$ and $s = 5$.

(b) The midpoint of $\overline{AC}$ is
$$\left(\frac{10+0}{2}, \frac{5+0}{2}\right) = \left(5, \frac{5}{2}\right).$$
The midpoint of $\overline{BD}$ is
$$\left(\frac{3+7}{2}, \frac{5+0}{2}\right) = \left(5, \frac{5}{2}\right).$$
Since these points are the same, the diagonals bisect each other.

32. (a) The distance from $(0, 0)$ to (x, y) is 3:
$\sqrt{(x-0)^2 + (y-0)^2} = 3$ or $x^2 + y^2 = 9$.

(b) A circle is by definition the set of all points that are a fixed distance (the radius) from a fixed point (the center). Therefore, this set of points is a circle of radius 3 with its center at the origin.

34. $\angle BCA \cong \angle DCE$ as vertical angles. $\angle ABC \cong \angle EDC$ as alternate interior angles when parallel lines are cut by a third line. Similarly, $\angle BAC \cong \angle DEC$. So by the AA similarity property (using any two of these facts), $\triangle ABC \sim \triangle EDC$.

36. $\angle CED \cong \angle AEB$ (since they are in fact the same angle). Since $\overline{CD}$ is parallel to $\overline{AB}$, the corresponding angles, $\angle CDE$ and $\angle ABE$, are congruent. Hence, by the AA similarity property, $\triangle CDE \sim \triangle ABE$. Since corresponding sides are proportional, we have
$$\frac{5}{7} = \frac{6-y}{6}$$
$30 = 42 - 7y$
$-12 = -7y$
$\frac{12}{7} = y$

JUST FOR FUN   **The Greek Cross—I (page 917)**

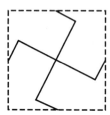

Problem Set 14.2 (page 929)

1. (a) $m = \frac{8-4}{3-1} = \frac{4}{2} = 2$; upward to the right

(c) $m = \frac{-7-(-3)}{-4-(-2)} = \frac{-4}{-2} = 2$; upward to the right

(e) $m = \frac{-5-(-2)}{-2-1} = \frac{-3}{-3} = 1$; upward to the right

2. $\frac{7-3}{4-b} = 2$
$4 = 2(4-b)$
$2 = 4-b$
$b = 2$

5. Substitute a for x and 3 for y:
$2a + 3 \cdot 3 = 18$
$2a = 9$
$a = \frac{9}{2}$

7. (a),(c)
For part (a), $3x + 5y = 12$.
If $x = -1$, then $-3 + 5y = 12$, or $y = 3$, so $(-1, 3)$ is on the line.
If $y = 0$, then $3x = 12$, or $x = 4$, so $(4, 0)$ is on the line.
For part (c), $5y - 3x = 15$.
If $x = 0$, then $5y = 15$, or $y = 3$, so $(0, 3)$ is on the line.
If $y = 0$, then $-3x = 15$, or $x = -5$, so $(-5, 0)$ is on the line.

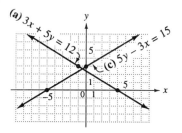

8. (a), (c), (e)
For part (a): $2y - 16 = 0$ means $y = 8$ (a horizontal line).
For part (c): $x + y = 2$ includes $(0, 2)$ and $(2, 0)$.
For part (e): $y = 3x + 4$ is the line with y-intercept 4 and slope 3.

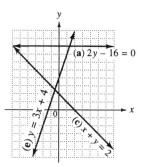

10. First, find the slope of $3x - 5y + 45 = 0$ by solving for y:
$0 = 3x - 5y + 45$
$5y = 3x + 45$
$y = \frac{3}{5}x + 9$

The slope is $\frac{3}{5}$.

(a) $7x + ky = 21$
$ky = -7x + 21$
$y = \frac{-7}{k}x + \frac{21}{k}$

The slope must be $\frac{3}{5}$:
$\frac{-7}{k} = \frac{3}{5}$
$3k = -35$
$k = -\frac{35}{3}$

(c) $y = kx + 5$
The slope must be $\frac{3}{5}$:
$k = \frac{3}{5}$

11. (a) $3x - 7y + 21 = 0$
$-7y = -3x - 21$
$y = \frac{3}{7}x + 3$
$m = \frac{3}{7}; b = 3$

(c) $y = 6$
$y = 0x + 6$
$m = 0; b = 6$

(e) $x = -5$
This is a vertical line. The slope is undefined and there is no y-intercept.

12. (a) Write each line in slope-intercept form so that the slopes can be compared.
$3x + 7y + 15 = 0 \qquad 6x + 15y + 31 = 0$
$7y = -3x - 15 \qquad 15y = -6x - 31$
$y = -\frac{3}{7}x - \frac{15}{7} \qquad y = -\frac{3}{5}x - \frac{31}{15}$

The slopes are $-\frac{3}{7}$ and $-\frac{3}{5}$. They are not the same, and their product is not -1. The lines are neither parallel nor perpendicular

(c) $x + 4y + 6 = 0 \qquad 8x = 2y + 13$
$4y = -x - 6 \qquad -2y = -8x + 13$
$y = -\frac{1}{4}x - \frac{3}{2} \qquad y = 4x - \frac{13}{2}$

The slopes are $-\frac{1}{4}$ and 4. Their product is -1. The lines are perpendicular.

13. $2x + ky + 6 = 0 \qquad 3x - 5y = 15$
$ky = -2x - 6 \qquad -5y = -3x + 15$
$y = -\frac{2}{k}x - \frac{6}{k} \qquad y = \frac{3}{5}x - 3$

The slopes are $-\frac{2}{k}$ and $\frac{3}{5}$.

(a) We require $-\frac{2}{k} = \frac{3}{5}$, so $3k = -10$.
Then $k = -\frac{10}{3}$.

14. (a) Find the slope-intercept form of each line:
$2x - 3y = 9 \qquad 4x - 4y = 16$
$-3y = -2x + 9 \qquad -4y = -4x + 16$
$y = \frac{2}{3}x - 3 \qquad y = x - 4$

The lines have different slopes, so they intersect. At the intersection point,
$\frac{2}{3}x - 3 = x - 4$
$1 = \frac{1}{3}x$
$x = 3$
$y = x - 4 = 3 - 4 = -1$
The lines intersect at $(3, -1)$.

16. (a) $\left(\frac{3+5}{2}, \frac{-5+9}{2}\right) = (4, 2)$

18. (a) Midpoint: $\left(\frac{3+(-1)}{2}, \frac{5+7}{2}\right) = (1, 6)$
Slope of $\overline{PQ}$: $m = \frac{7-5}{-1-3} = \frac{2}{-4} = -\frac{1}{2}$
The perpendicular bisector has slope 2 and passes through $(1, 6)$. Its equation in point-slope form is $y - 6 = 2(x - 1)$.

(c) Midpoint: $\left(\frac{3+4}{2}, \frac{-2+6}{2}\right) = \left(\frac{7}{2}, 2\right)$
Slope of $\overline{CD}$: $m = \frac{6-(-2)}{4-3} = \frac{8}{1} = 8$
The perpendicular bisector has slope $-\frac{1}{8}$ and passes through $\left(\frac{7}{2}, 2\right)$. Its equation in point-slope form is
$y - 2 = -\frac{1}{8}\left(x - \frac{7}{2}\right)$.

(e) Midpoint: $\left(\frac{-1+5}{2}, \frac{5+5}{2}\right) = (2, 5)$
Slope of $\overline{GH}$: $m = \frac{5-5}{5-(-1)} = \frac{0}{6} = 0$
The perpendicular bisector is a vertical line that passes through $(2, 5)$. Its equation is $x = 2$.

24. The shortest distance is along the line that is perpendicular to $3x - 2y = 6$ and contains $(1, 5)$. Since $3x - 2y = 6$ is equivalent to $-2y = -3x + 6$, or $y = \frac{3}{2}x - 3$, the perpendicular line has slope $-\frac{2}{3}$. Its equation in point-slope form is

$y - 5 = -\frac{2}{3}(x - 1)$, or $y = -\frac{2}{3}x + \frac{17}{3}$.

The lines $y = \frac{3}{2}x - 3$ and $y = -\frac{2}{3}x + \frac{17}{3}$ intersect at a point where

$\frac{3}{2}x - 3 = -\frac{2}{3}x + \frac{17}{3}$. Then

$9x - 18 = -4x + 34$

$13x = 52$

$x = 4$

$y = \frac{3}{2}x - 3 = \frac{3}{2}(4) - 3 = 3$

The lines intersect at (4, 3). The desired distance is the distance from (1, 5) to (4, 3), which is

$\sqrt{(4-1)^2 + (3-5)^2} = \sqrt{9+4} = \sqrt{13}$.

29. (a)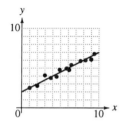

Many lines are possible. One line is $y - 6 = \frac{1}{2}(x - 8)$ or $y = \frac{1}{2}x + 2$. When $x = 15$, $y = \frac{1}{2}(15) + 2 = \frac{19}{2}$.

30. (a) Figure (iv) has only a vertical line of symmetry.

(c) Yes. Figure (ii) has 3 lines of symmetry.

32. No. The figure shown has 120° rotational symmetry, but no symmetry about a point.

JUST FOR FUN   **The Greek Cross–II (page 940)**

If each edge of the cross has length 1, then the area of the cross (and hence the area of the square) is 5. Therefore, the side of the square must be $\sqrt{5}$ units long.

Since $5 = 1^2 + 2^2$, this suggests that a side of the square can be formed from a cut that is the hypotenuse of a triangle whose legs have lengths 1 and 2. After some experimentation, we arrive at the solution shown.

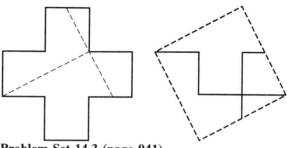

Problem Set 14.3 (page 941)

1. (a) Since $\overline{AB}$ is horizontal, $\overline{CD}$ must be horizontal, so $s = c$. Since $\overline{AD}$ is vertical, $\overline{BC}$ must be vertical, so $r = a$. Since AB and AD have the same length, $c = a$. Therefore, $r = a$, $s = a$, and $c = a$.

3. (a) Since $\overline{AD}$ is horizontal, $\overline{BC}$ must be horizontal, so $s = b$. Since AB and DC must have the same slope, $\frac{b-0}{a-0} = \frac{s-0}{r-c}$, or $\frac{b}{a} = \frac{b}{r-c}$.

Therefore, $r - c = a$, so $r = a + c$. Hence, $r = a + c$ and $s = b$.

5. (a) $(x - 0)^2 + (y - 0)^2 = 9^2$

Circle centered at (0, 0) with radius 9

(c) $(x - 0)^2 + (y - (-5))^2 = 0^2$

$\sqrt{(x-0)^2 + (y-(-5))^2} = 0$

This is the set of points whose distance from (0, –5) is 0—that is, the single point (0, –5).

(e) $(x - (-2))^2 + (y - (-3))^2 = 5^2$

Circle centered at (–2, –3) with radius 5

6. (a) $(x - 2)^2 + (y - (-3))^2 = 7^2$

Circle centered at (2, –3) with radius 7

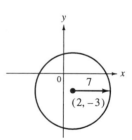

7. (a) $(x - 2)^2 + (y - 5)^2 = 3^2$

$(x - 2)^2 + (y - 5)^2 = 9$

(c) The point (–1, 2) is 2 units above the x-axis, so the radius is 2.
$(x-(-1))^2 + (y-2)^2 = 2^2$
$(x+1)^2 + (y-2)^2 = 4$

8. (a) Substitute 1 for x and 2 for y in the equation for the circle.
$(1+2)^2 + (2-3)^2 = 3^2 + (-1)^2$
$= 9 + 1 = 10$ so the coordinates of the point satisfy the equation.

10. (a) The center lies on the perpendicular bisector of any chord between two of the points. Thus, we can find the center of the circle by determining two such perpendicular bisectors and finding their intersection.
(7, 7) and (–1, 1):
Midpoint of chord:
$\left(\frac{7+(-1)}{2}, \frac{7+1}{2}\right) = (3, 4)$
Slope of chord: $\frac{1-7}{-1-7} = \frac{-6}{-8} = \frac{3}{4}$
The perpendicular bisector has slope $-\frac{4}{3}$ and passes through (3, 4). Its equation is
$y - 4 = -\frac{4}{3}(x - 3)$ or, equivalently,
$y = -\frac{4}{3}x + 8$.
(7, 7) and (6, 0):
Midpoint of chord:
$\left(\frac{7+6}{2}, \frac{7+0}{2}\right) = \left(\frac{13}{2}, \frac{7}{2}\right)$
Slope of chord: $\frac{0-7}{6-7} = \frac{-7}{-1} = 7$
The perpendicular bisector has slope $-\frac{1}{7}$ and passes through $\left(\frac{13}{2}, \frac{7}{2}\right)$. Its equation

is $y - \frac{7}{2} = -\frac{1}{7}\left(x - \frac{13}{2}\right)$ or, equivalently,
$y = -\frac{1}{7}x + \frac{31}{7}$.
These bisectors intersect at the point where
$-\frac{4}{3}x + 8 = -\frac{1}{7}x + \frac{31}{7}$
$-28x + 168 = -3x + 93$
$75 = 25x$
$x = 3$
$y = -\frac{4}{3}x + 8 = -\frac{4}{3}(3) + 8 = 4$
The center of the circle is (3, 4).
(*Note*: The perpendicular bisector of the chord from (–1, 1) to (6, 0) has the equation $y - \frac{1}{2} = 7\left(x - \frac{5}{2}\right)$ or, equivalently, $y = 7x - 17$. The point (3, 4) also satisfies this equation, as expected.)

11. A general parallelogram can be drawn as shown.

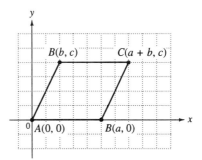

The midpoint of $\overline{AC}$ is $\left(\frac{a+b}{2}, \frac{c}{2}\right)$ and midpoint of $\overline{BD}$ is $\left(\frac{a+b}{2}, \frac{c}{2}\right)$. These points are the same, so the diagonals bisect each other.

14. (a) One method is given in the back of the textbook. An alternate method follows. Let P, Q, and R be the midpoints of $\overline{AB}$, $\overline{BC}$, and $\overline{CA}$, respectively. The coordinates of these points are $P\left(\frac{a+c}{2}, \frac{b+d}{2}\right)$, $Q\left(\frac{c+e}{2}, \frac{d+f}{2}\right)$, and $R\left(\frac{a+e}{2}, \frac{b+f}{2}\right)$.

Then G is on $\overline{AQ}$ because slope $\overline{AG} = \frac{\frac{b+d+f}{3} - b}{\frac{a+c+e}{3} - a} = \frac{-2b+d+f}{-2a+c+e}$ and

Slope $\overline{AQ} = \frac{\frac{d+f}{2} - a}{\frac{c+e}{2} - a} = \frac{-2b+d+f}{-2a+c+e}$.

G is on $\overline{BR}$ because slope $\overline{BG} = \dfrac{\frac{b+d+f}{3}-d}{\frac{a+c+e}{3}-c} = \dfrac{b-2d+f}{a-2c+e}$ and slope $\overline{BR} = \dfrac{\frac{b+f}{2}-d}{\frac{a+e}{2}-c} = \dfrac{b-2d+f}{a-2c+e}$.

G is on $\overline{CP}$ because slope $\overline{CG} = \dfrac{\frac{b+d+f}{3}-f}{\frac{a+c+e}{3}-e} = \dfrac{b+d-2f}{a+c-2e}$ and slope $\overline{CP} = \dfrac{\frac{b+d}{2}-f}{\frac{a+c}{2}-e} = \dfrac{b+d-2f}{a+c-2e}$.

Thus, G is on all three medians.

17. (a) The points are $P\left(\frac{2a}{2}, \frac{2a}{2}\right) = P(a, a)$, $Q\left(2a+\frac{2b+2c}{2}, \frac{2b+2c}{2}\right) = Q(2a+b+c, b+c)$,
$R\left(2a+2b+\frac{2c}{2}, -\frac{2c}{2}\right) = R(2a+2b+c, -c)$, and $S\left(\frac{2a+2b}{2}, -\frac{2a+2b}{2}\right) = S(a+b, -(a+b))$. Thus,
$PR = \sqrt{(2a+2b+c-a)^2 + (-c-a)^2} = \sqrt{(a+2b+c)^2 + (a+c)^2}$ and
$SQ = \sqrt{(2a+b+c-a-b)^2 + ((b+c)+(a+b))^2} = \sqrt{(a+c)^2 + (a+2b+c)^2}$.
These quantities are equal, so $\overline{PR} \cong \overline{SQ}$.

(b) Slope $\overline{PR} = \dfrac{-c-a}{(2a+2b+c)-a} = -\dfrac{a+c}{a+2b+c}$

Slope $\overline{SQ} = \dfrac{(b+c)-(-(a+b))}{(2a+b+c)-(a+b)} = \dfrac{a+2b+c}{a+c}$

Since the slope of $\overline{SQ}$ is the negative of the reciprocal of the slope of $\overline{PR}$, $\overline{PR} \perp \overline{SQ}$.

20. $(10-2) \cdot 180° = 1440°$

22. $\dfrac{1440°}{10} = 144°$

JUST FOR FUN   **Watering a Playfield (page 950)**

The plumber is right, as shown below. (If the field is square, he is "barely right.")

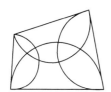

Problem Set 14.4 (page 953)

1. (a) $y = 2x - 3$

x	y
-3	-9
-2	-7
-1	-5
0	-3
1	-1
2	1
3	3

3. (a) Yes. This is the graph of a function because each vertical line cuts the graph just once.

5. (a) $4x - y < 8$. Graph the line $4x - y = 8$ as a dashed line, because equality is not allowed. Since $(0, 0)$ satisfies the inequality, shade the half-plane containing $(0, 0)$.

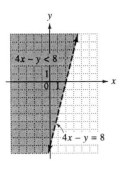

(c) $x + y \leq 4$

Graph the line $x + y = 4$ as a solid line, because equality is allowed. Since (0, 0) satisfies the inequality, shade the half-plane containing (0, 0).

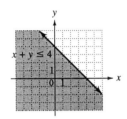

(e) $3x + 5y + 15 \geq 0$

Graph the line $3x + 5y + 15 = 0$ as a solid line, because equality is allowed. Since (0, 0) satisfies the inequality, shade the half-plane containing (0, 0).

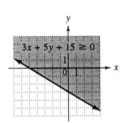

7. (a) $y = x^2$

x	y
−3	9
−2	4
−1	1
0	0
1	1
2	4
3	9

8. (a) $y = x^2 - 4x + 4$

x	y
−3	25
−2	16
−1	9
0	4
1	1
2	0
3	1

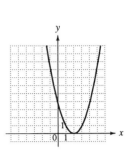

9. (a) $y = x^2 + 10x$

x	y
−8	−16
−7	−21
−6	−24
−5	−25
−4	−24
−3	−21
−2	−16
−1	−9
0	0
1	11
2	24

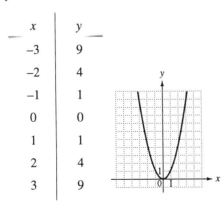

From the graph, the minimum value of $x^2 + 10x$ is −25. It occurs at $x = -5$.

12. (a) $y = x^2 + 2x$

x	y
−3	3
−2	0
−1	−1
0	0
1	3
2	8
3	15

(b) The line of symmetry is the vertical line $x = -1$.

13. (a) $y^2 = x + 4$

x	y
-5	—
-4	0
-3	±1
-2	$\pm\sqrt{2} \approx \pm 1.41$
-1	$\pm\sqrt{3} \approx \pm 1.73$
0	±2
5	±3

16. (a) $y = |x - 1|$

x	y
-3	4
-2	3
-1	2
0	1
1	0
2	1
3	2

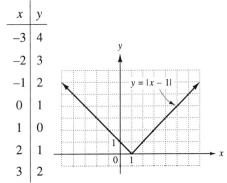

20. (a) Graph the function $y = 400 + 80x - x^2$. The maximum occurs at (40, 2000). The maximum number of shoes Acme can sell per week is 2000 shoes.

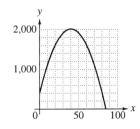

23. (a) Since the population doubles every minute, it was half-full one minute before noon—that is, at 11:59 A.M.

25. (a) In the diagram below D is a units from A. If D has coordinates (b, c), then by the Pythagorean Theorem $b^2 + c^2 = a^2$. Hence $c^2 = b^2 - a^2$ and $c = \sqrt{a^2 - b^2}$. Since $DC = a$, C has coordinates $\left(a + b, \sqrt{a^2 - b^2}\right)$. Thus, the midpoint of AC is $\left(\frac{a+b}{2}, \frac{\sqrt{a^2 - b^2}}{2}\right)$ and the midpoint of $\overline{BD}$ is also $\left(\frac{a+b}{2}, \frac{\sqrt{a^2 - b^2}}{2}\right)$. Since these points are the same, the diagonals bisect each other.

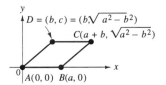

26. (a) The line is given in the slope-intercept form $y = mx + b$, where $m = -3$. The slope is –3.

28. The equation $2x - 7y = 14$ is equivalent to $y = \frac{2}{7}x - 2$, so it has slope $\frac{2}{7}$. In point-slope form, the equation of the line with slope $\frac{2}{7}$ that passes through (–2, –3) is $y + 3 = \frac{2}{7}(x + 2)$.

CLASSIC CONUNDRUM Square Inch Mysteries (page 957)

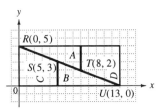

The square has area $8 \times 8 = 64$, and the rectangle has area $5 \times 13 = 65$. Place the rectangle on a coordinate system and label the points $R(0, 5)$, $S(5, 3)$, $T(8, 2)$ and $U(13, 0)$ as shown. Since

$$\text{slope } \overline{RS} = \frac{3-5}{5-0} = \frac{-2}{5} = -0.40$$

and

$$\text{slope } \overline{SU} = \frac{0-3}{13-5} = \frac{-3}{8} = -0.375,$$

R, S, and U do *not* lie on a straight line. Similarly,

$$\text{slope } \overline{RT} = \frac{2-5}{8-0} = -\frac{3}{8} = -0.375$$

and

$$\text{slope } \overline{TU} = \frac{0-2}{13-8} = \frac{-2}{5} = -0.40,$$

so R, T, and U do *not* lie on a straight line. Indeed, as the slopes show, $RTUS$ is a very thin parallelogram—a fact obscured by the heavy lines used in the diagram. It turns out that this parallelogram has area one and this accounts for the extra unit of area.

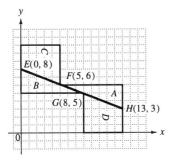

The "propeller" has area
$5 \times 6 + 5 \times 6 + 3 \times 1 = 63$. Place the "propeller" on a coordinate system and label the points $E(0, 8)$, $F(5, 6)$, $G(8, 5)$, and $H(13, 3)$ as shown. Here
$$\text{slope } \overline{EF} = \frac{6-8}{5-0} = -\frac{2}{5} = -0.40$$
and
$$\text{slope } \overline{FH} = \frac{3-6}{13-5} = \frac{-3}{8} = -0.375,$$
so E, F, and H do not lie on a straight line. Similarly,
$$\text{slope } \overline{EG} = \frac{5-8}{8-0} = \frac{-3}{8} = -0.375$$
and
$$\text{slope } \overline{GH} = \frac{3-5}{13-8} = \frac{-2}{5} = -0.40,$$
so E, G, and H also do not lie on a straight line. Indeed, as the slopes show, $EFHG$ forms a small parallelogram where the pieces of the puzzle overlap—a fact again obscured by the heavy lines used in making the drawing. It turns out that this parallelogram has area one, and this accounts for the missing one unit of area.

Chapter 14 Review Exercises (page 959)

1.

2. (a) Quadrant IV. Since $a > 0$, the point is to the right of the y-axis, and since $b < 0$, the point is below the x-axis.

   (b) Quadrant II. Since $a < 0$, the point is to the left of the y-axis, and since $b > 0$, the point is above the x-axis.

   (c) Quadrant I or Quadrant III. In these quadrants, x and y have the same sign.

   (d) Quadrant II or Quadrant IV. In these quadrants, x and y have opposite signs.

3. (a) $M(a, b) = M\left(\frac{1+3}{2}, \frac{5+(-1)}{2}\right) = M(2, 2)$. Hence $a = 2$ and $b = 2$.

   (b) $M(a, b) = M\left(\frac{3.2+1.4}{2}, \frac{1.7+(-1.5)}{2}\right) = M(2.3, 0.1)$. Hence $a = 2.3$ and $b = 0.1$.

   (c) $M(a, b) = M\left(\frac{0+(-5)}{2}, \frac{3+(-4)}{2}\right) = M(-2.5, -0.5)$. Hence $a = -2.5$ and $b = -0.5$.

   (d) $M(a, b) = M\left(\frac{0+(-2)}{2}, \frac{0+(-9)}{2}\right) = M(-1, -4.5)$. Hence $a = -1$ and $b = -4.5$.

4. (a) $\sqrt{(3-1)^2 + (-1-5)^2} = \sqrt{4+36}$
       $= \sqrt{40} = 2\sqrt{10}$

   (b) $\sqrt{(1.4-3.2)^2 + (-1.5-1.7)^2}$
       $= \sqrt{3.24 + 10.24}$
       $= \sqrt{13.48} \doteq 3.67$

   (c) $\sqrt{(-5-0)^2 + (-4-3)^2} = \sqrt{25+49}$
       $= \sqrt{74}$

   (d) $\sqrt{(-2-0)^2 + (-9-0)^2} = \sqrt{4+81} = \sqrt{85}$

5. $AB = \sqrt{(0-(-3))^2 + (5-1)^2}$
   $= \sqrt{9+16} = \sqrt{25} = 5$
   $AC = \sqrt{(1-(-3))^2 + (-2-1)^2}$
   $= \sqrt{16+9} = \sqrt{25} = 5$
   $BC = \sqrt{(1-0)^2 + (-2-5)^2} = \sqrt{1+49} = \sqrt{50}$
   $(AB)^2 + (AC)^2 = 5^2 + 5^2 = 50$
   $= \left(\sqrt{50}\right)^2 = (BC)^2$

   Thus, by the Pythagorean Theorem, the triangle is a right triangle.

6. $RS = \sqrt{(3-(-7))^2 + \left(0-\frac{9}{2}\right)^2}$
   $= \sqrt{100 + \frac{81}{4}} = \sqrt{\frac{481}{4}} = \frac{\sqrt{481}}{2}$

$$RT = \sqrt{(1-(-7))^2 + \left(-3-\frac{9}{2}\right)^2} = \sqrt{64 + \frac{225}{4}}$$
$$= \sqrt{\frac{481}{4}} = \frac{\sqrt{481}}{2}$$

$RS = RT$, so the triangle is isosceles.

7. (a) $m = \dfrac{-3-(-2)}{2-3} = \dfrac{-1}{-1} = 1$

   (b) $m = \dfrac{5-(-7)}{3-3} = \dfrac{12}{0}$ Undefined

   (c) $m = \dfrac{-7-(-5)}{1-(-2)} = \dfrac{-2}{3} = -\dfrac{2}{3}$

   (d) $m = \dfrac{-3-5}{2-2} = \dfrac{-8}{0}$ Undefined

8. $m = \dfrac{7-(-2)}{-2-c} = \dfrac{9}{-2-c}$

   If $\dfrac{9}{-2-c} = 5$, then $5(-2-c) = 9$, so
   $-2-c = \dfrac{9}{5}$. Then $c = -2 - \dfrac{9}{5} = -\dfrac{19}{5}$.

9. (a) Slope $\overleftrightarrow{CD}$ = Slope $\overleftrightarrow{EF}$
   $$\dfrac{b-5}{2-3} = \dfrac{-7-2}{2-(-1)}$$
   $$\dfrac{b-5}{-1} = \dfrac{-9}{3}$$
   $3(b-5) = 9$
   $b-5 = 3$
   $b = 8$

   (b) (slope $\overleftrightarrow{CD}$) · (slope $\overleftrightarrow{EF}$) = -1
   $$\left(\dfrac{b-5}{-1}\right) \cdot \left(\dfrac{-9}{3}\right) = -1$$
   $3(b-5) = -1$
   $b-5 = -\dfrac{1}{3}$
   $b = 5 - \dfrac{1}{3}$
   $b = \dfrac{14}{3}$

10. $R(-4, 5)$ is on the perpendicular bisector of PQ if, and only if, it is equidistant from P and Q—that is, if, and only if, $PR = QR$. Since
    $PR = \sqrt{(-4-0)^2 + (5-5)^2} = \sqrt{16+0} = 4$ and
    $QR = \sqrt{(-4-(-3))^2 + (5-1)^2}$
    $= \sqrt{1+16} = \sqrt{17} \doteq 4.12$, $PR \neq QR$.
    Therefore, $R(-4, 5)$ is *not* on the perpendicular bisector of $\overline{PQ}$.

11. Substitute 3 for x and b for y in the equation:
    $3 \cdot 3 - 5b = 17$
    $9 - 5b = 17$
    $-5b = 8$
    $b = -\dfrac{8}{5}$

12. Slope $\overline{PQ} = \dfrac{7-(-2)}{-4-3} = \dfrac{9}{-7} = -\dfrac{9}{7}$
    In point-slope form, the equation of the line is
    $y + 2 = -\dfrac{9}{7}(x-3)$ or, equivalently,
    $y - 7 = -\dfrac{9}{7}(x+4)$.

13. Use the slope-intercept form $y = mx + b$ where $m = \dfrac{3}{2}$ and $b = 10$. The equation is
    $y = \dfrac{3}{2}x + 10$.

14. Rewrite the equation in slope-intercept form by solving for y:
    $3x - 4y = 15$
    $-4y = -3x + 15$
    $y = \dfrac{3}{4}x - \dfrac{15}{4}$
    The slope is $\dfrac{3}{4}$.

15. The perpendicular line has slope $-\dfrac{4}{3}$. In point-slope form, its equation is
    $y - 1.5 = -\dfrac{4}{3}(x-5)$.

16. Solve each equation for y:
    $3x - 5y = 19$ $2x + 3y = 0$
    $-5y = -3x + 19$ $3y = -2x$
    $y = \dfrac{3}{5}x - \dfrac{19}{5}$ $y = -\dfrac{2}{3}x$

    At the point (x, y) where the lines intersect, we have:
    $\dfrac{3}{5}x - \dfrac{19}{5} = -\dfrac{2}{3}x$
    $9x - 57 = -10x$
    $19x = 57$
    $x = 3$
    $y = -\dfrac{2}{3}x = -\dfrac{2}{3} \cdot 3 = -2$
    The solution is $(3, -2)$; that is, $x = 3$ and $y = -2$.

17. (a) The coordinates of M, D, and E are $M(4, 3)$, $D(-3, 3)$, and $E(4, -4)$. Thus, $\overline{DM}$ is horizontal, $\overline{ME}$ is vertical and $\angle DME$ is a right angle.

(b) Slope $\overline{EA} = \dfrac{0-(-4)}{0-4} = -1$,

Slope $\overline{ED} = \dfrac{3-(-4)}{-3-4} = -1$.

Since the slopes are the same, E, A, and D are collinear.

18. (a) Since the coordinates of E, G, and H are $E(3, 3)$, $G(8, 2)$, and $H(5, -5)$.

$EG = \sqrt{(8-3)^2 + (2-3)^2}$
$= \sqrt{25+1} = \sqrt{26}$ and

$FH = \sqrt{(5-6)^2 + (-5-0)^2}$
$= \sqrt{1+25} = \sqrt{26}$.

Since $EG = FH$, we conclude $\overline{EG} \cong \overline{FH}$.

(b) Slope $\overleftrightarrow{EG} = \dfrac{2-3}{8-3} = \dfrac{-1}{5} = -\dfrac{1}{5}$

Slope $\overleftrightarrow{FH} = \dfrac{-5-0}{5-6} = \dfrac{-5}{-1} = 5$

Since the product of these slopes is -1, it follows that $\overleftrightarrow{EG}$ and $\overleftrightarrow{FH}$ are perpendicular.

19. (a) Since the coordinates of E, F, G, and H are $E(2, 2)$, $F(5, 1)$, $G(9, 3)$, and $H(6, -6)$.

$EG = \sqrt{(9-2)^2 + (3-2)^2}$
$= \sqrt{49+1} = \sqrt{50}$

$FH = \sqrt{(6-5)^2 + (-6-1)^2}$
$= \sqrt{1+49} = \sqrt{50}$

Since $EG = FH$, we conclude $\overline{EG} \cong \overline{FH}$.

(b) Slope $\overleftrightarrow{EG} = \dfrac{3-2}{9-2} = \dfrac{1}{7}$

Slope $\overleftrightarrow{FH} = \dfrac{-6-1}{6-5} = \dfrac{-7}{1} = -7$

Since the product of these slopes is -1, it follows that $\overleftrightarrow{EG}$ and $\overleftrightarrow{FH}$ are perpendicular.

20. (a) Since the coordinates of M, D, and E are $M(a, b)$, $D(-b, b)$, and $E(a, -a)$, $\overline{MD}$ is horizontal and $\overline{ME}$ is vertical. Thus, $\angle EMD$ is a right angle and $\triangle EMD$ is a right triangle.

(b) Slope $\overline{AE} = \dfrac{-a-0}{a-0} = \dfrac{-a}{a} = -1$

Slope $\overline{AD} = \dfrac{b-0}{-b-0} = \dfrac{b}{-b} = -1$

Since these slopes are equal, it follows that E, A, and D are collinear.

21. (a) The coordinates of E, G, and H are $E(a, a)$, $G(2a + b, b)$ and $H(a + b, -a - b)$. Therefore,

$EG = \sqrt{((2a+b)-a)^2 + (b-a)^2} = \sqrt{(a+b)^2 + (b-a)^2}$
$= \sqrt{(a^2 + 2ab + b^2) + (b^2 - 2ab + a^2)}$
$= \sqrt{2a^2 + 2b^2}$

and

$FH = \sqrt{((a+b)-2a)^2 + ((-a-b)-0)^2}$
$= \sqrt{(b-a)^2 + (a+b)^2}$
$= \sqrt{(b^2 - 2ab + a^2) + (a^2 + 2ab + b^2)}$
$= \sqrt{2a^2 + 2b^2}$

Since $EG = FH$, it follows that $\overline{EG} \cong \overline{FH}$.

(b) Slope $\overleftrightarrow{EG} = \dfrac{b-a}{(2a+b)-a} = \dfrac{b-a}{b+a}$

Slope $\overleftrightarrow{FH} = \dfrac{(-a-b)-0}{(a+b)-2a} = \dfrac{-a-b}{b-a} = -\dfrac{b+a}{b-a}$

Since (slope $\overleftrightarrow{EG}$) $\cdot$ (slope $\overleftrightarrow{FH}$) = $\left(\dfrac{b-a}{b+a}\right) \cdot \left(-\dfrac{b+a}{b-a}\right) = -1$, $\overleftrightarrow{EG}$ and $\overleftrightarrow{FH}$ are perpendicular.

22. (a) The coordinates of E, F, G, and H are $E(a, a)$, $F(2a + b, b)$, $G(2a + 2b + c, c)$, and $H(a + b + c, -(a + b + c))$. Therefore,

$$EG = \sqrt{((2a+2b+c)-a)^2 + (c-a)^2} = \sqrt{(a+2b+c)^2 + (c-a)^2} \text{ and}$$

$$FH = \sqrt{((a+b+c)-(2a+b))^2 + (-(a+b+c)-b)^2}$$
$$= \sqrt{(-a+c)^2 + (-a-2b-c)^2} = \sqrt{(c-a)^2 + (a+2b+c)^2}.$$

Since $EG = FH$, it follows that $\overline{EG} \cong \overline{FH}$.

(b) Slope $\overleftrightarrow{EG} = \dfrac{c-a}{(2a+2b+c)-a} = \dfrac{c-a}{a+2b+c}$

Slope $\overleftrightarrow{FH} = \dfrac{-(a+b+c)-b}{(a+b+c)-(2a+b)} = \dfrac{-a-2b-c}{-a+c} = -\dfrac{a+2b+c}{c-a}$

Since (slope $\overleftrightarrow{EG}$) · (slope $\overleftrightarrow{FH}$) $= \left(\dfrac{c-a}{a+2b+c}\right) \cdot \left(-\dfrac{a+2b+c}{c-a}\right) = -1$, $\overleftrightarrow{EG}$ and $\overleftrightarrow{FH}$ are perpendicular.

Chapter 14 Test (page 961)

1. For parallelogram $ABCD$, the point $C(10, 5)$ is easily found by drawing a sketch. Hence, $r = 10$ and $s = 5$.

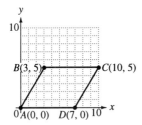

   Note: Actually, there are two other possibilities for point C. They are $C(-4, 5)$, which would produce parallelogram $ACBD$, and $C(4, -5)$, which would produce parallelogram $ABDC$.

2. The desired point is $\left(\dfrac{3}{4} \cdot (-8) + \dfrac{1}{4} \cdot 4, \dfrac{3}{4} \cdot 0 + \dfrac{1}{4} \cdot 12\right) = (-6 + 1, 0 + 3) = (-5, 3)$.

3. The midpoint of $\overline{AB}$ is $\left(\dfrac{2+7}{2}, \dfrac{5+4}{2}\right) = \left(\dfrac{9}{2}, \dfrac{9}{2}\right)$. The slope of $\overline{AB}$ is $\dfrac{4-5}{7-2} = \dfrac{-1}{5} = -\dfrac{1}{5}$, so the slope of the perpendicular bisector is 5. In point-slope form, the equation of the perpendicular bisector is
$y - \dfrac{9}{2} = 5\left(x - \dfrac{9}{2}\right)$.

4. $RS = \sqrt{(7-(-3))^2 + (10-2)^2} = \sqrt{100 + 64} = \sqrt{164}$
   $RT = \sqrt{(1-(-3))^2 + (-3-2)^2} = \sqrt{16 + 25} = \sqrt{41}$
   $ST = \sqrt{(1-7)^2 + (-3-10)^2} = \sqrt{36 + 169} = \sqrt{205}$
   Therefore, $(RS)^2 + (RT)^2 = 164 + 41 = 205 = (ST)^2$.
   By the Pythagorean theorem, this means that $\triangle RST$ is a right triangle.

5. (slope $\overline{PQ}$) = (slope $\overline{RS}$)

$$\frac{4-(-5)}{-3-2} = \frac{1-r}{4-2}$$

$$\frac{9}{-5} = \frac{1-r}{2}$$

$$2 \cdot 9 = -5(1-r)$$

$$-\frac{18}{5} = 1-r$$

$$r = 1 + \frac{18}{5} = \frac{23}{5}$$

6. Slope $\overline{PQ} = \frac{5-2}{-1-4} = \frac{3}{-5} = -\frac{3}{5}$

   In point-slope form, the equation is
   $y - 2 = -\frac{3}{5}(x-4)$ or, equivalently,
   $y - 5 = -\frac{3}{5}(x+1)$.

7. Rewrite the equation in slope-intercept form by solving for y:
   $5x + 2y = 19$
   $$2y = -5x + 19$$
   $$y = -\frac{5}{2}x + \frac{19}{2}$$
   The slope is $-\frac{5}{2}$.

8. $(x-(-2))^2 + (y-5)^2 = 4^2$
   $(x+2)^2 + (y-5)^2 = 16$

9. Solve each equation for y:
   $2x - 3y = 9$ $3x - 5y = 14$
   $-3y = -2x + 9$ $-5y = -3x + 14$
   $y = \frac{2}{3}x - 3$ $y = \frac{3}{5}x - \frac{14}{5}$

   At the point where the two lines intersect, we have:
   $$\frac{2}{3}x - 3 = \frac{3}{5}x - \frac{14}{5}$$
   $$10x - 45 = 9x - 42$$
   $$x = 3$$
   $y = \frac{2}{3}x - 3 = \frac{2}{3} \cdot 3 - 3 = 2 - 3 = -1$
   The simultaneous solution is $x = 3$ and $y = -1$, that is, $(3, -1)$.

10.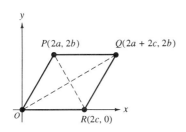

Any parallelogram can be placed on a coordinate axes as shown and with vertices with coordinates as indicated with a, b, and c all positive. The midpoint of $\overline{PR}$ is $\left(\frac{2a+2c}{2}, \frac{2b+0}{2}\right) = (a+c, b)$. The midpoint of $\overline{OQ}$ is $\left(\frac{0+(2a+2c)}{2}, \frac{0+2b}{2}\right) = (a+c, b)$. Since these midpoints are the same, the diagonals bisect each other.

11. $\$5000 \cdot (1.07)^5 \doteq \$7012.76$

12. (a) $y = 2x^2 + 4x$

x	y
-3	6
-2	0
-1	-2
0	0
1	6
2	16
3	30

    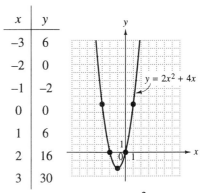

    (b) The minimum value of $2x^2 + 4x$ is -2. It occurs when $x = -1$.

13. $2x - 6y \leq 12$
    Since equality is allowed, graph $2x - 6y = 12$ as a solid line. Since $(0, 0)$ satisfies the inequality, shade the half-plane that includes $(0, 0)$.

    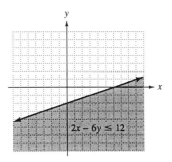

Appendix B

1. (a) Statement, since it is a false declarative sentence.

   (c) Not a statement, since without a specified value for x, the equation is neither true nor false. (Exception: the equation could be *stipulated* to be true; this would constitute a specification of x's value.)

2. (a) "For some whole number x, $x + 5 = 9$" is true, since the equation is true for the whole number $x = 4$.

   (c) "For some whole number x, $x \cdot 1 = x \cdot 2$" is true, since the equation is true for the whole number $x = 0$.

4. (a) $3 \cdot 1 \neq 3$

   (c) $2 \not< 1$ or, equivalently, $2 \geq 1$.

   (e) It is false that some rectangles are not squares or, equivalently, All rectangles are squares.

   (g) Not all squares have a right angle or, equivalently, Some squares do not have a right angle.

5. (a) $\sim p$ ("$\sim$" means "not.")

   (c) $\sim p \rightarrow \sim q$ ("$\rightarrow$" means "if ... then.")

6. (a) True, since p is true and the operation is "at."

   (c) False, since s is false and the operation is "and."

7. (a) True, since $2 + 2 = 4$ is true and the operation is "or."

8. (a) "It is false that I own a calculator and drive a car" becomes "I do not own a calculator or I do not drive a car."

10. (a) Converse: If r is a moosh, then r is a dweeble.
    Contrapositive: If r is not a moosh, then r is not a dweeble.

11. (a) The inverse is logically equivalent to its own contrapositive; $\sim p \rightarrow \sim q$ is equivalent to $\sim\sim q \rightarrow \sim\sim p$, which by the law of double negation (applied to each part) is equivalent to $q \rightarrow p$, which is the converse of $p \rightarrow q$.

    (b) If r is not a dweeble, then r is not a moosh. If $s \cdot t = s \cdot u$, then $t = u$.

13. (a) Let p be "n^2 is odd" and q be "n is odd."
    Proof of $q \rightarrow p$: Suppose q, i.e., n is odd. Then for some whole number i, $n = 2i + 1$, and $n^2 = (2i+1)^2 = 4i^2 + 4i + 1 = 2(2i^2 + 2i) + 1$, which is also odd. So p is true, which establishes $q \rightarrow p$.
    Proof of $p \rightarrow q$: Suppose $\sim q$, i.e., n is even. Then for some whole number j, $n = 2j$, and $n^2 = (2j)^2 = 4j^2 = 2(2j^2)$, which is also even. So $\sim p$ is true, which establishes $\sim q \rightarrow \sim p$ and, by taking the contrapositive, $p \rightarrow q$.
    Hence $p \rightarrow q$.

    (b) For whole numbers, being odd is equivalent to being not even. This allows one to break the result of Example B.8 into two conditionals, take the contrapositive of each, replace "not even" with "odd," and recombine the resulting conditionals.

14. (a) Valid; an instance of contraposition, since for whole numbers, "$x + y + z$ is even" is equivalent to "$x + y + z$ is not odd."

    (c) Invalid, since the premises are true but the conclusion is false.

    (e) Invalid, since $x = 8$ is a counterexample. (Both premises are true, but the conclusion is false.)

15. (a) Fallacy of the converse. (Also, the fact that nothing bad happened so far does not entail that nothing bad will happen in the future.)

16. (a) q, since "p or q" and "$\sim p$" can both be true only if q is true.

18. (a) True, by the law of detachment applied to $r \rightarrow s$ and r.

    (c) True, by syllogism applied to $p \rightarrow q$ and $q \rightarrow r$.

    (e) Cannot be decided, since s is true, but q's status is unknown (see a and b above).

    (g) False, since f is.

19. (a) Valid. Let p be "It rains," q be "It won't snow," and r be "it is below freezing." Then the argument is a syllogism with the second premise contraposed. (Note, however, that treating "It is above freezing" as "it is not below freezing" assumes that either "above" or "below," but not both, covers the case where the temperature is *at* freezing.)

21. (a)

$1 = 2^0 \cdot 1$	$13 = 2^0 \cdot 13$
$2 = 2^1 \cdot 1$	$14 = 2^1 \cdot 7$
$3 = 2^0 \cdot 3$	$15 = 2^0 \cdot 15$
$4 = 2^2 \cdot 1$	$16 = 2^4 \cdot 1$
$5 = 2^0 \cdot 5$	$17 = 2^0 \cdot 17$
$6 = 2^1 \cdot 3$	$18 = 2^1 \cdot 9$
$7 = 2^0 \cdot 7$	$19 = 2^0 \cdot 19$
$8 = 2^3 \cdot 1$	$20 = 2^2 \cdot 5$
$9 = 2^0 \cdot 9$	$21 = 2^0 \cdot 21$
$10 = 2^1 \cdot 5$	$22 = 2^1 \cdot 11$
$11 = 2^0 \cdot 11$	$23 = 2^0 \cdot 23$
$12 = 2^2 \cdot 3$	$24 = 2^3 \cdot 3$

(b) No; 16 has degree 4.

(c) h and k are odd numbers, and $h < k$. Let $2u$ be an even number between h and k. Then $m < 2^d \cdot 2u < n$, and $2^d \cdot 2u = 2^{d+1} u$, so the degree of $2^d \cdot 2u$ is at least $d + 1$.

(d) Suppose two numbers m and n in a consecutive finite list have the same degree of evenness d, and that this is the largest degree in the list. By the result in part (c) there is a number between m and n with a higher degree of evenness, and since the list is consecutive, this number is on the list—which contradicts the assumption. So the assumption cannot be true.

23. (a) If the tiling is possible, the sum of totals from each tile is a sum of multiples of 3, and must itself be a multiple of 3. But the sum must also, by inspection be 62—which is not a multiple of 3. So the tiling is not possible.

25. (a) The game always ends in a tie.

(b) The number of scalene and equilateral triangles is the same. By placing the eight-vertex configuration at an arbitrary position in the plane, this proves an interesting theorem: Let each point of the plane be colored red or blue, and let T be any scalene triangle of sides a, b, and c. Then there is a triangle in the plane congruent to T with vertices of one color if, and only if, there is an equilateral triangle of side a, b, or c with vertices of one color.

27. False. If the last digit of n^2 is 1, then the last digit of n is either 1 or 9. If d is the tens digit of n, squaring $10d + 1$ and $10d + 9$ yields $100d^2 + 20d + 1$ and $100d^2 + 180d + 81 = 100d^2 + (180d + 80) + 1$. In either case, the tens digit is even (since $2d$ and $18d + 8$ are both even). In particular, the tens digit is not 1. So no perfect square can end in 11.

28. (a) $\sim(\sim p) \equiv p$

(c) $p \wedge \sim p \equiv F$

(e) $(p \to q) \to ((p \wedge \sim q) \equiv F)$

29. (a) The circuit is open (false) just when $\sim p$ is open (false—i.e., p is true) and q is open (false).

Appendix C

Appendix C (page 999)

1. Answers will vary. The sample solution below draws LOGO.

   ```
   to LOGO
       pu lt 90 fd 150 rt 90 pd
       fd 80 bk 80 rt 90 fd 40
       pu fd 30 lt 90 pd
       repeat 2 [ fd 80 rt 90 fd 40 rt 90 ]
       pu rt 90 fd 70 lt 90 pd
       fd 80 rt 90 fd 40 rt 90 fd 30
       bk 30 lt 90 bk 40 lt 90 bk 80
       rt 90 fd 40 lt 90 fd 30 lt 90 fd 10 bk 20
       fd 10 rt 90 bk 30
       pu rt 90 fd 30 lt 90 pd
       repeat 2 [ fd 80 rt 90 fd 40 rt 90 ]
   end
   ```

6. Answers will vary.

   ```
   to right.triangle :leg1 :leg2
     fd :leg1 rt 90
     fd :leg2
     home
   end
   ```

   right.triangle 60 90

7. (a)
   ```
   to rectangle1
       repeat 2 [ fd 50 rt 90 fd 80 rt 90 ]
   end
   ```

   (b)
   ```
   to rectangle :length :width
       repeat 2 [ fd :length rt 90 fd :width rt 90 ]
   end
   ```

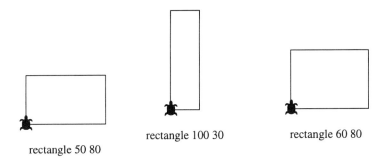

rectangle 50 80   rectangle 100 30   rectangle 60 80

8. (a) Note that $180° - 38° = 142°$. Therefore, the procedure may be accomplished as follows.

```
to parallelogram1
    repeat 2 [ fd 40 rt 142 fd 70 rt 38 ]
end
```

(b) ```
to parallelogram :side1 :side2 :angle
    repeat 2 [fd :side1 rt 180 - :angle fd :side2 rt :angle]
```

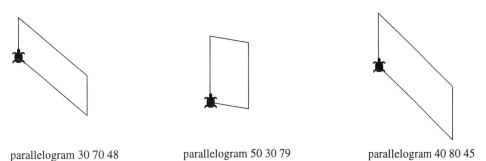

parallelogram 30 70 48   parallelogram 50 30 79   parallelogram 40 80 45

9. (a) One method is given below. Another way would be to use the procedure from problem 8(b) with equal values of **:side1** and **:side2**.
```
to rhombus1
    repeat 2 [fd 50 rt 128 fd 50 rt 52]
end
```

(b) ```
to rhombus :side :angle
    repeat 2 [ fd :side rt 180 - :angle fd :side rt :angle ]
end
```

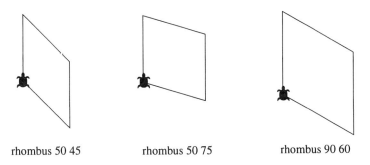

rhombus 50 45   rhombus 50 75   rhombus 90 60

Appendix C: Calvin T. Long and Duane W. DeTemple